Trends and Opportunities of Rapid Prototyping Technologies

Edited by Răzvan Păcurar

Published in London, United Kingdom

Trends and Opportunities of Rapid Prototyping Technologies
http://dx.doi.org/10.5772/intechopen.97962
Edited by Răzvan Păcurar

Contributors
Suchana Akter Jahan, Hazim El-Mounayri, Hakovirta Marko, Aggarwal Salonika, María Teresa Flores-Arias, Bastián Carnero, Carmen Bao-Varela, Ana Isabel Gómez-Varela, Denis Kotarski, Petar Piljek, Matija Krznar, Nino Krznar, Răzvan Păcurar

First published in London, United Kingdom, 2022 by IntechOpen
IntechOpen is the global imprint of INTECHOPEN LIMITED, registered in England and Wales, registration number: 11086078, 5 Princes Gate Court, London, SW7 2QJ, United Kingdom

British Library Cataloguing-in-Publication Data
A catalogue record for this book is available from the British Library

Additional hard and PDF copies can be obtained from orders@intechopen.com

Trends and Opportunities of Rapid Prototyping Technologies
Edited by Răzvan Păcurar
p. cm.
Print ISBN 978-1-80355-726-7
Online ISBN 978-1-80355-727-4
eBook (PDF) ISBN 978-1-80355-728-1

Meet the editor

Răzvan Păcurar is an associate professor in the Department of Manufacturing Engineering, Faculty of Industrial Engineering, Robotics and Production Management, Technical University of Cluj-Napoca, Romania. He has a PhD in Industrial Engineering and is a leading specialist in additive manufacturing. He has more than ninety journal articles to his credit and has presented at different international scientific conferences all over the world. He is the author of more than ten books in additive manufacturing, unconventional technologies, and computer-aided design. He has also edited three books and been involved in several national and international research projects as research team member and project leader. Dr. Păcurar is a member of the Manufacturing Engineering University Association of Romania and a journal reviewer.

Contents

Preface

Rapid prototyping technologies, also known as 3D printing or additive manufacturing technologies, are continuously developing due to their potential applications in the industrial and medical sectors. New types of materials (plastic, metallic, ceramic, etc.) are being conceived every year and testing of these materials using 3D printing technologies is of great interest among researchers worldwide.

As such, this book focuses primarily on the trends, challenges, and opportunities of these new modern manufacturing technologies. It also examines rapid prototyping (3D printing) in general with a focus on environmental challenges, recycling, energy usage and cost, and the need for standardization and certification in this field. The book discusses barriers and challenges of 3D printing methods, as well as customization of different methods for certain applications in the industrial sector.

Chapters emphasize at a very practical level how computer-aided design is required to be approached in close correlation with the process parameters and structural properties of the realized parts to be made by 3D printing. They also address how the designing process of customized parts must be considered in concordance with the particularities of the rapid prototyping (3D printing) technologies to be used for producing these parts.

This book is a useful resource for researchers, Ph.D. students, professors, engineers, and other readers interested in 3D printing.

Răzvan Păcurar
Department of Manufacturing Engineering,
Faculty of Industrial Engineering, Robotics and Production Management,
Technical University of Cluj-Napoca,
Cluj-Napoca, Romania

Chapter 1

Introductory Chapter: Rapid Prototyping – Trends and Opportunities

Răzvan Păcurar

1. Introduction

Since their advent, Rapid Prototyping technologies, nowadays known also as Additive Manufacturing or 3D printing methods became widely used in different applications that are used both, in industrial and medical domains [1, 2]. The benefits and advantages of using such modern manufacturing methods based on adding layers of materials, instead of removing (subtracting) material are multiple, consisting in the efficiency in producing parts more rapidly and with a minimum waste of material in the end [3]. Besides the advantages of these methods, there are also some disadvantages that consist in the performances that are reached using these types of technologies in terms of accuracy, roughness, and density in close correlation with the type of materials that are used in the additive manufacturing (Rapid Prototyping) methods and complexity of the parts that are required to be manufactured [4].

2. General presentation on the main Rapid Prototyping and Rapid Tooling methods

The manufacturing processes (most of them) are still quite expensive and the range of materials that can be used is still quite limited, even if in the last year's lot of progress has been made and quite new types of materials that are suitable to be used on the 3D printing technological methods occurred on the market [5].

Diversity of Rapid Prototyping methods is really impressive. The methods that are using molten plastic material that is heated up and extruded through a nozzle are the most spread and developed nowadays on the market [6]. Manufacturing of parts made of resin material by Vat polymerization (curing of the material using UV light) or manufacturing of different parts by Material Jetting or Binder Jetting, methods that are based on the use of different binders or fusing agents that are being sintered or cured by light are also 3D printing technological variants that are used on a larger scale in this field in the world [7]. Powder bed fusion or direct energy deposition methods are widely used to produce different types of metallic parts made by sintering or welding metallic powder grains in order to materialize one part made by 3D printing technologies [8, 9]. The parts can be made from a different range of metallic materials using these methods, starting with light materials like aluminum alloys or nickel alloys and ending with materials that have much higher strength like titanium alloys, tool steel alloys, etc. [10–13]. In the last year's lot of researches were reported

IntechOpen

in this field regarding new types of metallic materials that were developed and tested, like copper-alloys, gold, tungsten, carbide, or diamond 3D printing [14].

Rapid Tooling methods, such as Vacuum Casting, Metal Spraying, Investment casting, and other similar technologies were also quite well developed in the last years for producing different types of molds that were used to conceive, develop, and test new types of materials that are not suitable to be used by 3D printing technologies [15–17].

3. Trends, challenges and existing opportunities in the 3D printing domain

There are different types of materials that are still very difficult to be printed, like PMMA, Hydroxyapatite, or chitosan if we are referring to the medical domain or different types of materials that are easier to be made by casting instead of 3D printing, such as different types of Irons, so, therefore, these methods are considered to be also very important in this field in close correlation with the 3D printing technologies [18]. Rapid Tooling methods are using 3D printed parts as master models for the realized molds and, therefore, it is one real challenge to produce master models made of different types of materials that are suited to the Rapid Tooling method that is used for producing the molds in the end [19]. New types of materials mixed with new types of 3D printing or rapid tooling methods that are more and more used in the last years are really encouraging, are very promising, and represent one strong proof that there is still one wide open room in this domain, as there are still lot of things to be investigated in this field, in order to provide one quick response to the medical or industrial sector needs [20].

Life cycle of the products became very short, developing and personalizing of products are highly demanding also, need for standardization in the field of 3D printing is also one very important challenge as new types of additive. Manufacturing (3D printing) methods are still under development and are launched every year on the market [21].

The use of subtractive technological methods (like CNC cutting) in combination with additive manufacturing technologies in so-called "Hybrid manufacturing" technologies represents one of the most important trends on the market and provides lot of research topics for the researchers all over the world [22]. Whether we are referring to metallic, ceramic, or plastic materials, in principle, the trends, challenges, or opportunities are the same in the field of Rapid Prototyping technologies. Using of robots integrated with these methods, integration of "in-line" quality control systems in the manufacturing chain jointly with 3D printing methods, application of different types of heat treatment or coating methods in the rapid development process of a new product, and use of "digitalization" or "optimization" procedures or software programs correlated with the new developed 3D printing methods are just few of the most important examples of the opportunities in this field, both, for industrial or medical sectors [23–26].

Topological optimization or bionic design methods are quite well correlated with the additive manufacturing technologies, being also quite well integrated with specific software programs that are usually used to optimize the shape and redesign some components in order to decrease their weight, this trend being highly seen in the industrial sector (aerospace or automotive), but also in the medical sector, when such methods are used to enhance the biocompatibility of different medical implants, by designing the models with specific lattice structures in which bioactive materials can be inserted or integrated [27, 28]. Biomimetic structures, which refer actually to structures that imitate nature or biomimetic materials that can be used on 3D printing equipment are on top interest in the medical domain as well [29]. Bioengineering, biomedical, or bio 3D printing are new domains in which progress that has been made in the last years is really impressive and remarkable. Vessels, tissues, personalized bone structures, or even organs can be now realized or produced using 3D printing or 4D printing methods [30, 31].

Last, but not least, possibility of using multi-materials in the process of 3D printing and the possibility of using composite materials that are suitable to be used by 3D printing have enlarged the domains of applicability of these methods even more, by providing access to different applications, such as building constructions, textile and fashion, and consumer goods industry, etc. [32–34].

4. Conclusions

In the context of the trends and challenges that were presented in this chapter and based on the opportunities that exist in the Rapid Prototyping (3D printing) domain as were presented in this chapter, the current book aims to present few of the most important results that were reached in the 3D printing domain by researchers coming from different prestigious universities all over the world. The character of the researches presented in the book is inter and transdisciplinary, providing anyone who is looking to develop new researches in the field of 3D printing (PhD students, researchers, engineers, etc.) a good starting point for this purpose. The applications presented in this book are very large, covering a wide spectrum of 3D printing methods that were successfully used for the developing, testing, and producing different components that were further successfully used in the industrial and medical fields in the last years.

Author details

Răzvan Păcurar
Technical University of Cluj-Napoca, Romania

*Address all correspondence to: razvan.pacurar@tcm.utcluj.ro

References

[1] Salmi M. Additive manufacturing processes in medical applications. Materials (Basel). 2021;**14**(1):191. DOI: 10.3390/ma14010191

[2] Matthias S, Matthias G, Max L-G, Benjamin H, George S, Marius L, et al. Evaluating the use of additive manufacturing in industry applications. Procedia CIRP. 2019;**81**:19-23. DOI: 10.1016/j.procir.2019.03.004

[3] Tang Y, Mak K, Zhao Y. A framework to reduce product environmental impact through design optimization for additive manufacturing. Journal of Cleaner Production. 2016;**137**:1560-1572. DOI: 10.1016/j.jclepro.2016.06.037

[4] Addamo G, Peverini OA, Paonessa F, Virone G, Tascone R, Manfredi D, et al. Additive Manufacturing Technology for High Performances Feed Horn. 2018 IEEE International Symposium on Antennas and Propagation & USNC/URSI National Radio Science Meeting, 2018:1881-1882. DOI: 10.1109/APUSNCURSINRSM.2018.8608753

[5] Li N, Huang S, Zhang G, Qin RY, Liu W, Xiong H, et al. Progress in additive manufacturing on new materials. Journal of Materials Science & Technology. 2018;**35**:242-269. DOI: 10.1016/j.jmst.2018.09.002

[6] Kristiawan R, Imaduddin F, Ariawan D, Sabino U, Arifin Z. A review on the fused deposition modeling (FDM) 3D printing: Filament processing, materials, and printing parameters. Open Engineering. 2021;**11**:639-649. DOI: 10.1515/eng-2021-0063

[7] Bachmann J, Obst P, Knorr L, Schmoelzer S, Fruhmann G, Witt G, et al. Cavity vat photopolymerisation for additive manufacturing of polymer-composite 3D objects. Communications Materials. 2021;**2**:107. DOI: 10.1038/s43246-021-00211-5

[8] Ladani L, Sadeghilaridjani M. Review of powder bed fusion additive manufacturing for metals. Metals. 2021;**11**:1391. DOI: 10.3390/met11091391

[9] Dass A, Moridi A. State of the art in directed energy deposition: From additive manufacturing to materials design. Coatings. 2019;**9**:418. DOI: 10.3390/coatings9070418

[10] Qbau N, Nam ND, Hien NT, Ca NX. Development of light weight high strength aluminum alloy for selective laser melting. Journal of Materials Research and Technology. 2020;**9**:14075-14081. DOI: 10.1016/j.jmrt.2020.09.088

[11] Yap CY, Tan H, Du Z, Chua C, Dong Z. Selective laser melting of nickel powder. Rapid Prototyping Journal. 2017;**23**:750-757. DOI: 10.1108/RPJ-01-2016-0006

[12] Singla A, Banerjee M, Sharma A, Singh J, Bansal A, Gupta M, et al. Selective laser melting of Ti6Al4V alloy: Process parameters, defects and post-treatments. Journal of Manufacturing Processes. 2021;**64**:161-187. DOI: 10.1016/j.jmapro.2021.01.009

[13] Katancik M, Mirzababaei S, Ghayoor M, Pasebani S. Selective laser melting and tempering of H13 tool steel for rapid tooling applications. Journal of Alloys and Compounds. 2020;**849**:156319. DOI: 10.1016/j.jallcom.2020.156319

[14] Kurzynowski T, Gruber K, Chlebus E. The Use of Selective Laser Melting as a Method of New Materials Development, Lecture Notes in Mechanical

Engineering book series (LNME). 2019. DOI: 10.1007/978-3-030-04975-1_47

[15] Senthil J, Prabhahar M, Thiagarajan C, Sekar P, Lakshmanan R. Studies on performance and process improvement of implementing novel vacuum process for new age castings. Materials Today: Proceedings. 2020;**33**:813-819. DOI: 10.1016/j.matpr.2020.06.269

[16] Wortmann M, Frese N. Industrial-scale vacuum casting with silicone molds: A review. Applied Research. 2022;**1**:1-17. DOI: 10.1002/appl.202100012

[17] Gupta G, Tyagi R, Kumar R, Sunil M, Rahul JS, Verma S. Review on Thermal Spray Coating Methods and Property of Different Types of Metal-Based Coatings. Advances in Engineering Materials. 2021; DOI: 10.1007/978-981-33-6029-7_40

[18] Anketa J, Ikshita C, Ishika W, Ankush R, Mir I, Haq U. 3D printing—A review of processes, materials and applications in industry 4.0. Sustainable Operations and Computers. 2022;**3**:33-42. DOI: 10.1016/j.susoc.2021.09.004

[19] Böhme A, Nemak D, Schütze F, Lietzau K, Wolf E, Foitzik A. Adaption of 3D printing for rapid tooling. Materials Science Forum. 2021;**1016**:280-285. DOI: 10.4028/www.scientific.net/MSF.1016.280

[20] Dipak M, Hajare T, Gajbhiye S. Additive manufacturing (3D printing): Recent progress on advancement of materials and challenges. Materials Today: Proceedings. 2022;**58**(2):736-743. DOI: 10.1016/j.matpr.2022.02.391

[21] Rajat K, Harrsh K, Dubey S, Lokhande P. A review for advancements in standardization for additive manufacturing. Materials Today: Proceedings. 2022;**50**(5):1983-1990. DOI: 10.1016/j.matpr.2021.09.333

[22] Pragana JPM, Sampaio RFV, Bragança IMF, Silva CMA, Martins PAF. Hybrid metal additive manufacturing: A state–of–the-art review. Advances in Industrial and Manufacturing Engineering. 2021;**2**:100032. DOI: 10.1016/j.aime.2021.100032

[23] Werner J, Aburaia M, Raschendorfer A, Lackner M. MeshSlicer: A 3D-printing software for printing 3D-models with a 6-axis industrial robot. Procedia CIRP. 2021;**99**:110-115. DOI: 10.1016/j.procir.2021.03.018

[24] Vandone A, Baraldo S, Anastassiou D, Marchetti A, Valente A. 3D vision system integration on additive manufacturing machine for in-line part inspection. Procedia CIRP. 2020;**95**:72-77. DOI: 10.1016/j.procir.2020.01.191

[25] Phanden RK, Aditya SV, Sheokand A, Goyal KK, Gahlot P, Jacso A. A state-of-the-art review on implementation of digital twin in additive manufacturing to monitor and control parts quality. Materials Today: Proceedings. 2022;**56**(1):88-93. DOI: 10.1016/j.matpr.2021.12.217

[26] Sundar SS, Sundarlingam P, Laxmikant DS, Harshavardhana N, Ullengala S, Krishna PS. Numerical simulation process parameter optimization in metal additive manufacturing for getting better quality of products. Materials Today: Proceedings. online 13 May 2022. DOI: 10.1016/j.matpr.2022.04.455

[27] Prathyusha ALR, Babu GR. A review on additive manufacturing and topology optimization process for weight reduction studies in various industrial applications. Materials Today: Proceedings. 2022;**62**:109-117. DOI: 10.1016/j.matpr.2022.02.604

[28] Li Z, Xuan P, Lihua J, Cheng H, Jianxun S, Fei X, et al. Bionic design

and 3D printing of porous titanium alloy scaffolds for bone tissue repair. Composites Part B: Engineering. 2019;**162**:154-161. DOI: 10.1016/j.compositesb.2018.10.094

[29] Anton d. P., Adewumi J. B, Suvash C. P, Biranchi P, Jonathan P. T, Chris B. Biomimicry for 3D concrete printing: A review and perspective. Additive Manufacturing. 2021;**38**:101823. DOI: 10.1016/j.addma.2020.101823

[30] Pradeep PV, Paul L. Review on novel biomaterials and innovative 3D printing techniques in biomedical applications. Materials Today: Proceedings. 2022;**58**(1):96-103. DOI: 10.1016/j.matpr.2022.01.072

[31] Ayushi U, Kumar V, Sanjay M, Nand JK. Biomimetic 4D printed materials: A state-of-the-art review on concepts, opportunities, and challenges. Materials Today: Proceedings. 2021;**47**(11):3313-3319. DOI: 10.1016/j.matpr.2021.07.148

[32] Pajonk A, Prieto A, Blum U, Knaack U. Multi-material additive manufacturing in architecture and construction: A review. Journal of Building Engineering. 2022;**45**:103603. DOI: 10.1016/j.jobe.2021.103603

[33] Xiao Y-Q, Kan C-W. Review on development and application of 3D-printing Technology in Textile and Fashion Design. Coatings. 2022;**12**:267. DOI: 10.3390/coatings12020267

[34] Praveena BA, Lokesh N, Buradi A, Santhosh N, Praveena BL, Vignesh R. A comprehensive review of emerging additive manufacturing (3D printing technology): Methods, materials, applications, challenges, trends and future potential. Materials Today: Proceedings. 2022;**52**(3):1309-1313. DOI: 10.1016/j.matpr.2021.11.059

Chapter 2

New Industrial Sustainable Growth: 3D and 4D Printing

Aggarwal Salonika and Hakovirta Marko

Abstract

The commercial or industrial applications of 3D printing or additive manufacturing are continuously increasing in diverse areas mainly in rapid prototyping. 3D printing has become part of a novel industrial growth area where simplification of assembly, waste minimization, and mass customization are important, such as aerospace, orthopedic and medical research, defense, and jewelry. There has been continuous growth or improvement in additive manufacturing, which includes the type of materials used, metamaterials, and advancements in the printers or the software. 3D printing has explored the areas where materials have been manufactured which are several times lightweight, high strength compared to traditional parts, and also resulted in a reduction in CO_2 emissions. Biodegradability and sustainability are the major concern for any industry. The price of conventional thermoplastic filaments is one of the main sources of revenue and profitability for the industry. In addition to its relatively high price, some of the concerns in its wide use are the moisture resistance and VOC emissions, including iso-butanol and methyl-methacrylate (MMA) during 3D printing. These emissions cause voids in the structure which compromises the mechanical strength of the 3D-printed objects. Additives have been added with thermoplastics, such as diatoms and biodegradable materials, such as ceramics, biomaterials, graphene, carbon fibers, binders for metals, sand, and plaster to reduce the cost and VOC emissions. The cost of these additives is relatively less than the thermoplastic filaments. There has been tremendous innovative growth in the field of additive manufacturing, including solutions such as 3D-printed houses and titanium drones. The addition of additives opens the new potential applications in new arising technology, especially in robotics like behavior, mechanisms respond to user demands which are known as 4D printing where new dimension has been added to 3D printing. It is a process where a 3D-printed object transforms itself into another structure over the influence of external energy input, such as temperature, light, or other environmental stimuli. 4D printing is simply referred to as 3D printing transforming over time. 4D printing is an all-new emerging area in the field of additive manufacturing which has diverse applications in biomedical, defense, robotics, etc.

Keywords: 3D printing, biodegradability, thermoplastic filaments, additives, cost, biomaterials, 4D printing

1. Introduction

The process of joining materials layer upon layer from 3D digital model data or Computer-Aided Design (CAD) model is known as additive manufacturing (AM) or 3D printing as per International Organization for Standardization (ISO)/American Society for Testing and Materials (ASTM) 52900:2015 standard [1]. 3D printing has a long history of development for using it in the rapid prototyping of products for manufacturing since the 1980s. This development has since then led to also accessibility to the public. These developments started when Chuck Hull of 3D System Corp. filed their patent for a stereolithographic process eventually evolving into a 3D-printing technology boom [2]. Today 3D printer is priced as low as $100 [3] and is therefore accessible to the general public. Recent advances in 3D printing include, for example, the manufacturing of biomaterials for biomedical applications, such as tissue engineering. With recent advancements in the 3D printers, the industrial printers can build as small layers as 16 μm and thus creating a major milestone for biomedical applications [4]. 3D-printing technology can be used in various forms of materials printing, including fused deposition modeling (FDM), stereolithography (SLA), selective laser melting (SLM), and electron beam melting (EBM). The most used techniques are stereolithography and fused deposition modeling [5].

The International Organization for Standardization (ISO)/American Society for Testing and Materials (ASTM) 52900:2015 has classified the additive manufacturing (AM) process into seven categories (**Figure 1**) [5, 6].

There are several benefits to using 3D printing, such as [5, 7]:

1. Design to component translation.

2. Greater customization.

3. Manufacturing of complex, flexible, or lightweight components with no additional cost.

4. Potential of zero-waste manufacturing.

5. On-demand manufacturing.

6. Excellent scalability.

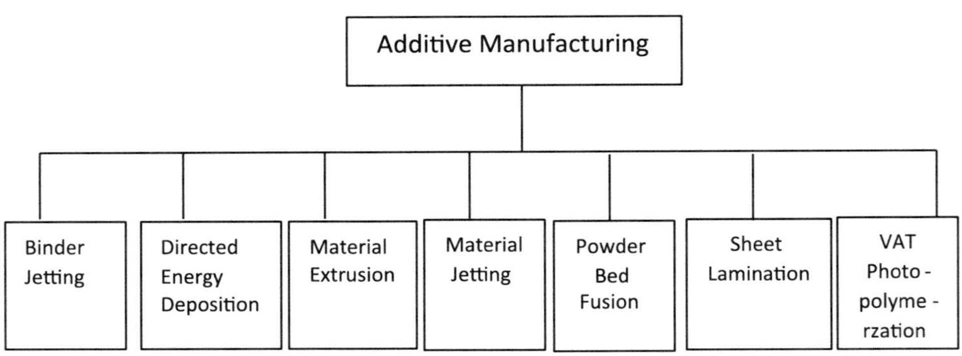

Figure 1.
Additive manufacturing processes.

Although the 3D-printing industry is rapidly growing, there have been several economic, social, and environmental challenges that need to be addressed, such as recycling of materials, energy usage, organic compounds emission, high cost of raw materials, and standards and certifications [6]. The lack of printing material [8] and the high cost of thermoplastic polymers add to the barrier to the industrialization of 3D-printing technologies [9]. The market growth potential is considerable for 3D-printing as it is estimated that the filament market will be worth $ 6.6 billion by 2026 [10]. One concern for the advancement of 3D printing other than the high cost of raw material is the emission of volatile organic compounds (VOC), including iso-butanol and methyl-methacrylate [11]. To address the abovementioned economic and environmental concerns, there has been a new advancement in the additive manufacturing process which includes the addition of additives can such as diatoms [10] and biodegradable materials, such as ceramics, biomaterials, graphene, carbon fibers, binders for metals, sand, and plaster [12]. The cost of these additives is relatively much less than the thermoplastic filaments. In addition, there are added benefits including included in the addition of additives, such as improved moisture resistance that may slow down the process of decomposition of the filament material and may potentially open up other innovative functional possibilities, such as immobilization of chemical sensors and bacteria and virus-killing agents for novel biomedical applications.

In general, the structures fabricated with 3D printing either using single or multiple materials are intrinsically static, hence 3D printing cannot meet the demands where dynamic materials applications are needed including, for example, hygromorph biocomposites [13], adaptive wind turbines [14], active biocomposites [15], and self-folding microgrippers [16]. This addition of a new dimension to 3D printing has started a new era of printing known as 4D printing and includes novel materials compositions, additives, and chemical functionalization.

2. Barriers and challenges of 3D printing

There are several challenges associated with manufacturing or scaling up of 3D printing mentioned as follows [17]:

Personal customization vs mass manufacturing: Currently AM technology is suitable for customized products, and low-volume production whereas for high-volume production, still injection molding is being industrially used because of the high-cycle time of additive manufacturing. However, there are industrial sections where high-cycle time is balanced by high demand for customized and complex geometry products, reduction in material waste, and opportunity to merge parts, such as GE fuels nozzle and customized earphones by Ownphones.

Material heterogeneity and structural consistency: For most of the industrial additive manufacturing processes, single materials are used which do not show any heterogeneousness and structural discrepancy at the interface but with the more advancement in the 3D printing, multiple materials are being used to accomplish more complex geometry of the products. Different materials have different behaviors, properties, and functionalities which leads to anisotropic mechanical properties of 3D-printed products due to the interlayer bonding deficiencies which eventually limits the variety of materials used for additive manufacturing. More research needs to be done for analyzing the uncertain behavior of multiple materials and CAD or computer software needs to be redesigned for the adoption of multiple materials in additive manufacturing.

Building scalability vs layer resolution: The two parameters scale of building parts and layer resolution are inversely proportional to each other in 3D printing which means that with the increase in the size or layer thickness of 3D-printed parts, the layer resolution decreases which results in the deprived surface quality because of layer stair-stepping effect. Typically layer resolution of 0.1 mm and layer thickness of 25 mm is used for commercial implications of 3D printing. There has been recent advancement with inkjet printing where the researchers have managed to get the layer thickness as low as 0.012 mm but at the cost of high-building time [18]. However, researchers have managed to produce products from nanoscale to macro to large scale using hybrid AM processes, coating, and post-processing subtractive machining.

Intellectual Property and AM standardization: To ensure the consistency of additive manufactured product, there is a certain need for the standardization of AM process, material, machine, and file format. ASTM has started to approve AM material standardization for several processes but there is a long way ahead for AM machine manufacturers and researchers. The open access to the downloadable file has opened a new challenge to protect the intellectual property right of researchers and commercial machine manufacturers. There is a need for planned and diverse patent filing for additive manufacturing processes which should include material, process, CAD design, machining, and post-processing modifications.

3. Applications

Earlier 3D printing or additive manufacturing was normally used for rapid prototyping only but in the current scenario, 3D printing has already established a large pool of diverse applications, for example, in manufacturing, sociocultural, food, and biomedical sectors. There is a wide range of applications from nano to macro to large scale for 3D printing (**Figure 2**).

Figure 2.
Range of applications of 3D printing.

4. Recent advancements

There are several types of advancements are done recently to increase the efficiency of the additive manufacturing process, such as materials advancement, process advancement, and post-processing advancement.

4.1 Material advancements

There are several challenges associated with 3D printing, such as emission of volatile organic compounds, creation of voids, and high cost of thermoplastic polymers. To avoid all these issues, recent advancements have been done which include the use of fillers, such as carbon fibers, nanofibers, graphite, and diatomaceous earth [9, 10]. Carbon nanotube/polylactic acid composites (CNT/PLA) and multi-walled carbon nanotube/polylactic acid composites (MWCNT/PLA) with strong mechanical properties are being explored in microelectronics [19]. The smaller particles sizes are used in composites to produce stiffness and high density in the printed products, such as hydroxyapatite-reinforced polyethylene/polyamide composites (HA-PE/PA) [20]. Carbon black/polyamide 12 (CA/PA12) composites were fabricated which enhance the mechanical, thermal, and electrical properties of printed products [21]. Nanomaterial composites, such as nanosilica/polyamide, nanoclay/polyamide, and graphite nanoplatelets/polyamide composites, have also been fabricated leading to improved mechanical properties [9]. These composites can be used for multiple applications, such as biomedical applications, because of the high surface area of fillers present (**Figure 3**) [10].

Metamaterials: Metamaterials are another innovation provided by 3D printing in material science. The functional metamaterials are considered a complex machines working. For instance, there is a material developed by the Lawrence Livermore National Labs where it gets shrunk when heated up rather than expanding. USC Viterbi has created a metamaterial that can manipulate sound through the magnetic field. Metamaterial has the iron particles in its lattice structure and in the presence of a magnetic field the structure gets deformed into one which blocks the sound rather than passing them through. Similarly, researchers at Boston University have developed the metamaterial by 3d printing the plastic coil used in metamaterial which has improved the MRI scan quality and speed and also blocks 94% sound [22].

Titanium Drones: Titomic has recently used their Titomic Kinetic Fusion Process to 3D print the titanium drone by mechanically fusing the titanium powder. This innovative 3D-printing method allows to use of different materials or alloys in a single prototype and eliminated the problems associated with traditional manufacturing, such as welds (**Figure 4**) [22].

4.2 Pioneering bionic 3D printing

There is an innovative advancement that mimics the living organism's organic cellular structure and bone growth. The world's largest 3D-printed airplane cabin

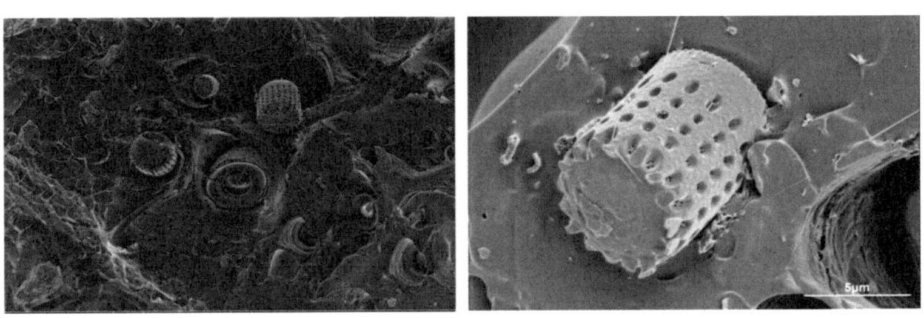

Figure 3.
3D-printed Diatoms in the PLA matrix (original work).

Figure 4.
3D-printed titanium firefighting drone [23].

component with a "bionic partition" which separates the passenger cabin from the galley has been divulged by Autodesk and Airbus. This design has made the partition very light with a 45% reduction in weight compared to traditional designs but still very strong. It has been estimated that this design would save 465,000 metric tons of CO_2 emissions per year. This new bionic partition used the second-generation alloy of scandium, aluminum, and magnesium named "Scalmalloy" created by the 3D-printing expertise of Airbus subsidiary "APWorks" (**Figure 5**) [24].

Similarly, Airbus has collaborated with Materialise to produce the 3D-printed bionic spacer panel using FDM and Materialise's post-production processes which made the panel 15% light in weight compared to traditional panels (**Figure 6**) [25, 26].

Stratasys has been 3D printing more Airbus cabin components for years now [27]. Airbus A350 XWB was decided to be manufactured by 3D printing (**Figure 7**) [28].

4.3 Product advancement

4.3.1 3D-Printed house

Alquist 3D has printed the first-ever 3D-printed house in the US which was assembled in 22 hours. The printer head was connected to the tube through which the traditional concrete was being pumped. Alquist 3D has teamed up with the nonprofit organization known as "Habitat for Humanity" where they will be providing homes to the people in need. Alquist 3D has claimed that the 3D-printed houses are 10–15% less in cost compared to traditional house building. It has saved the manpower also as according to Alquist 3D, only 3–4 humans were required to operate the printer [29]. This was not the first time 3D-printed houses have been built. In France, 3D-printed houses were built and Europe's first 3D-printed house was built in 22 days which was later shortened to 3 days. In Dubai, there have been 3D-printed offices have been built. According to the Dubai government, it has saved them almost 50% of the total cost [30]. Initially, 3D printing was used only for prototyping the construction but now 3D printing has been used for constructing the whole buildings.

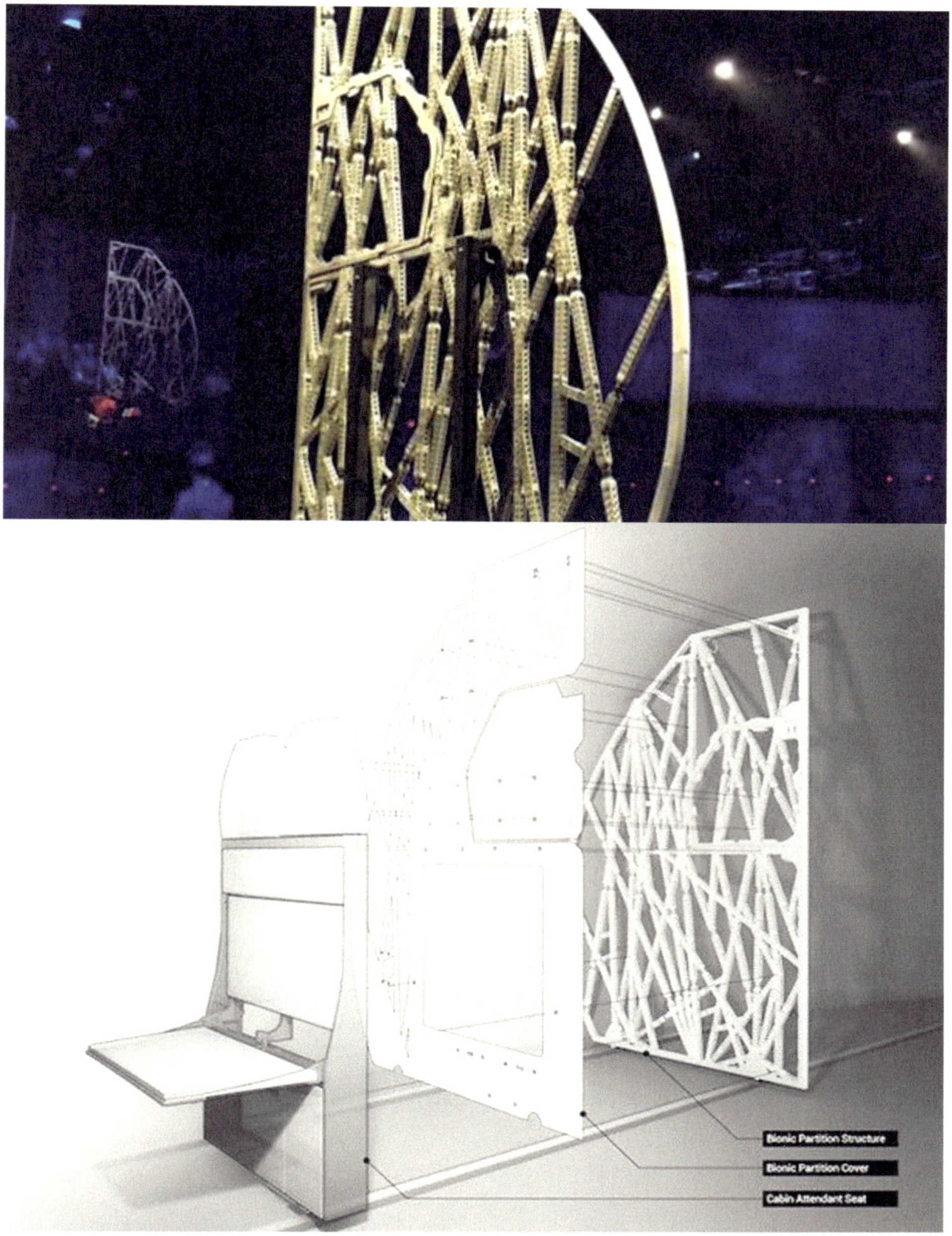

Figure 5.
Airbus 3D printed bionic partition cabin [Source: Airbus].

4.3.2 Porsche's revolutionizing product development

Porsche has used 3D-printing technology to produce 3D-printed pistons, spare parts, and sports seats. Porsche has developed the lightweight, better thermal resistance, high-performance pistons for the twin-turbo boxer engine of the 911GT2 RS

Figure 6.
3D-printed finished spacer panel, [Source: Materialise].

Figure 7.
Airbus 3D metal-printed bionic titanium bracket [Source: Airbus].

model leading to a 30-horsepower gain. This process used the laser printing or laser metal fusion process in collaboration with MAHLE & TRUMPF which uses the high-precision machine, TruPrint 3000 with a 500-Watt fiber laser and high-purity metal special aluminum alloy powder which melted to print 1200 layers ending into the desired shape (**Figure 8**) [31, 32].

Porsche has been manufacturing spare parts using selective laser melting since 2018 but recently, Porsche has started manufacturing personalized bodyform full-bucket sports seats for Porsche 911 and 718. Porsche has also invested in 3D-printing specialist INTAMSYS (**Figure 9**).

Figure 8.
Pistons of the twin-turbo boxer engine of 911GT2 RS [Source: Porsche AG].

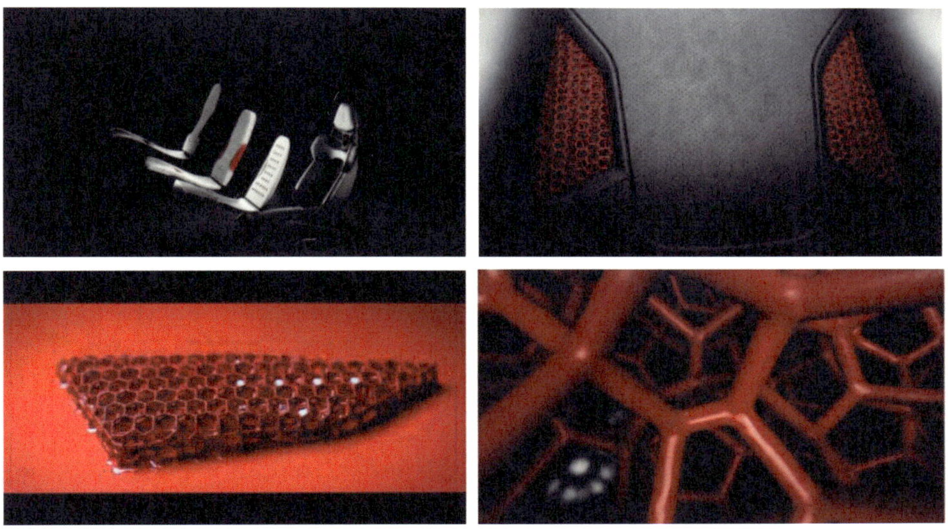

Figure 9.
3D-Printed bodyform full-bucket sports seats [Source: Porsche AG].

Porsche has also produced its first complete housing for its electric drive using the additive laser fusion process which has opened the possibilities for 3D printing in the highly stressed electric sports cars sector (**Figure 10**) [33].

4.4 Post-processing advancement

Achieving Good surface quality: As it was mentioned in section 3, there is a discrepancy with the surface quality whenever multiple materials are used because of the layer stair-stepping effect, but there have been numerous advancements done in this field to achieve the good surface quality such as mask-image-projection-based stereolithography (MIP-SL) which is a hybrid AM stereolithography process. In the

Figure 10.
Prototype for small series production [Source: Porsche AG].

MIP-SL process, Digital Micromirror Device (DMD) is used instead of a laser, unlike stereolithography. DMD is an electromechanical device that can control ~1 million small mirrors simultaneously to turn on or off a pixel at over 5kHz [18] which results in the projection of masked images on a surface area. MIP-SL process is faster than the SLA process. MIP-SL was developed to spread the liquid resin on the smooth ultrathin layer surface which basically used the meniscus equilibrium method for building the smooth up-curved surfaces [18]. There have been several approaches used to achieve the good surface quality, such as blasting, sanding, chemical finishing, which has differentiation influence on the surface quality. There is another post-processing step to achieve good surface quality is the coating method on the 3D-printed parts. For instance, there is a chemical coating XTC-3D was used because it was lost cost, easy to work on any 3D-printing surfaces (FDM, SLS, SLA) which showed that coating has filled up the gap between the layers and improved the final finished surface quality of the product [34].

Protecting Intellectual Property and AM standardization: To protect the intellectual property rights of any process, material, and file information, there has been advancement where information is either encrypted or embedded into the structure domain such as infraStructs which literally means below the structures where tags including all the information are embedded inside the digitally fabricated structure which can be read only by Terahertz (TZ) imaging system. It is not visible outside the surface and can be read only by advanced imaging systems [35]. Similarly, there is an advancement where watermarks can be extracted from the 3D prints by changing the original 3D print mesh [36, 37].

5. 4D Printing

4D printing or smart printing has a unique basic characteristic that differentiates it from the static 3D-printing structures; 4D-printing materials are dynamic and able to have functionality [8]. The well-used definition describing the 4D-printing states "It is the evolution of a 3D printed structure either in shape, property, and functionality when it is exposed to external factors such as light [38], heat [39], pH [40], and water [41]".

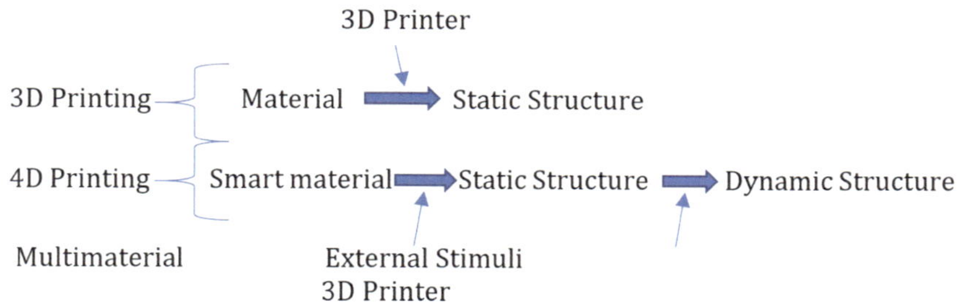

Figure 11.
3D vs 4D printing.

4D printing can be defined as the best combination of a smart material, a 3D printer, and a well-programmed automated design (**Figure 11**) [8].

There are five factors that influence the 4D printing which are the additive manufacturing process, feedstock material, stimuli, interaction mechanism, and modeling [42].

According to F. Momeni and J. Ni, there are three laws that define the shape-changing behavior of 4D-printed objects [43]. The first law states that "all the shapes changing behaviors such as curling, twisting, coiling, bending, etc. of multi-material 4D structures are due to the relative expansion between active and passive materials."

Types of materials	Examples	References
Responsive toward moisture: Hydrogels	Hydrogels respond to moisture or water and can expand up to 200% of their original volume. Sustainable materials, such as cellulose, can be used as hydrogel printing ink compatible with various types of printers	[8, 14, 40]
Responsive toward light: photo-responsive	Chromophore (photosensitive) materials are inserted into smart material for which light acts as an indirect stimulus because light generates the heat which eventually changes the shape of the material.	[8, 37]
Responsive toward temperature: thermo-responsive	Temperature (heating or cooling) is used as an external stimulus either to change the shape of material – shape change effect (SCE) or to transform the deformed shape into the original shape – shape memory effect (SME). SMEs can be polymers, metals, ceramics, alloys, and gels. These smart materials are used in biomedical applications such as orthodontics, physiotherapy, orthopedics, surgeries, etc.	[8, 38]
Materials responsive toward pH	Polyelectrolytes are used as smart material which changes their shape as the pH changes with the release or gain of protons. It has found applications in biocatalysts, valves, actuators, drug delivery, etc.	[8, 39]
Materials responsive toward the electric field	An electric field is also the indirect stimulus that produces the heat and causes the change in shape. For instance, origami using polypyrrole	[8]
Materials responsive toward the magnetic field	Smart materials change their shape in the presence of a magnetic field. Magnetic nanoparticles are incorporated into hydrogels which respond in the presence of a magnetic field	[8, 41, 42]
Piezoelectric materials	The charge is produced with mechanical stress which eventually causes the deformation.	[8, 41, 42]

Table 1.
Types of materials used in 4D printing.

Figure 12.
4D printed metamaterials reconfigurable object [48, 49].

The second law states that "there are four physical factors behind the shape changing ability of all multi-material 4D structures i.e., mass diffusion, thermal expansion, molecular transformation, and organic growth."

The third law states that the "time-dependent shape-morphing behavior of nearly all multi-material 4D printed structures is governed by two "types" of time constants" (**Table 1**).

There are revolutionary applications associated with the 4D printing, such as biomedical applications of 4D printing in drug delivery, organ regeneration and transplantation, and tissue fabrication [44]. 4D-printed structures have great potential in soft robotics because of their capability to deform, adjust to environmental changes, and flexibility [8]. 4D-printed structures with smart materials can be used as self-evolving structures [45, 46], active origami structures [47], self-sustainable satellite manufacturing parts [8], sensors responsive toward moisture, temperature, pH, magnetic energy, etc. [8]. Despite diverse applications, 4D printing needs more research and development, especially in scaling it up. Commercializing the 4D printing is troublesome because of the high production cost, installation cost and material used and availability. Multi-materials printers could be a possible solution but need furthermore research (**Figure 12**).

6. Conclusion

Additive manufacturing was invented in the 1940s and it has developed a lot with innovative inventions since then. The different additive manufacturing process techniques have specific peculiarities and the disadvantage of one technique can lead to the innovation of a new technique. The development of different types of printers has enabled the AM to use different types of materials which include plastics, metals, and ceramics. New improvements in AM techniques allow the high filler loading in thermoplastic composites.

3D printing has diverse applications include for instance food, fashion, biomedical, health, aerospace, and cultural heritage preservation. 3D printing helps the consumer to customize the product as per their requirements. There are a few challenges that need to be addressed, such as emission of volatile organic compounds, creation of voids, high cost of thermoplastic polymers, and weak mechanical strength, of printed structures. To overcome these challenges composites with fillers have been fabricated such as carbon nanotube/polylactic acid composites, nanosilica/polyamide composites, and carbon black/polyamide composites which have increased the mechanical, electrical, and thermal properties of the composites.

Despite highly diverse applications of 3D printing and new advancements in 3D printing, there are still a few challenges that restrict the usage of 3D printing on a commercial scale. These include the resistance and adaptability of 3D-printed material's properties and structures against the change in environmental factors, such as temperature, electric energy, and pH.

4D printing is basically the combination of a 3D printer, smart material, and well-designed programming that allows the 3D-printed object to change its shape, properties, and functionality with time. 4D-printed objects change or modify against environmental conditions. These materials can be responsive to heat, water, pH, electric energy, and magnetic field. 4D printing has increased the number of application areas for additive manufacturing and thus expanded to include aviation, self-sustaining material, sensors, active materials, and bioprinting.

There has been a tremendous amount of technological advancement and research on 3D and 4D printing, and its applications. New advancements have been, however, the commercialization and implementation at a larger stage are still in progress and therefore more research and development are needed. Importantly more sustainable materials need to be explored due to the environmental risks associated with some of the materials and techniques used. The potential to create solutions to some of the most challenging product development needs in various industries using 3D- and 4D-printing technologies remain high. These developments are many times related to niche products that cannot be manufactured otherwise.

Author details

Aggarwal Salonika and Hakovirta Marko*
North Carolina State University, Raleigh, North Carolina, USA

*Address all correspondence to: mjhakovi@ncsu.edu

References

[1] Thakar CM, Parkhe SS, Jain A, et al. 3d Printing: Basic principles and applications. Materialstoday: Proceedings. 2022;**51**(1):842-849

[2] Hull CW. Apparatus for production of three-dimensional ojects. Stereolithography. 1986;**4**:575

[3] Alexandrea P. The Top Cheap 3D Printers on the Market. *The 3Dnatives*; 2022

[4] Chia HN, Wu BM. Recent advances in 3D printing of biomaterials. Journal of Biological Engineering. 2015;**9**(1):1-14. DOI: 10.1186/s13036-015-0001-4

[5] Tofail SAM, Koumoulos EP, Bandyopadhyay A, Bose S, O'Donoghue L, Charitidis C. Additive manufacturing: Scientific and technological challenges, market uptake and opportunities. Materials Today. 2018;**21**(1):22-37. DOI: 10.1016/j.mattod.2017.07.001

[6] Additive Manufacturing: Strategic Research Agenda (2014). Vol. 1192014. pp. 460-493

[7] Li S, Xin Y, et al. Design for additive manufacturing from a force-flow perspective. Materials & Design. 2021;**204**:109664

[8] Ahmed A, Arya S, Gupta V, Furukawa H, Khosla A. 4D printing: Fundamentals, materials, applications and challenges. Polymer. 2021;**228**:123926

[9] Valino AD, Dizon JRC, Espera AH, Chen Q, Messman J, Advincula RC. Advances in 3D printing of thermoplastic polymer composites and nanocomposites. Progress in Polymer Science. 2019;**98**:101162. DOI: 10.1016/j.progpolymsci.2019.101162

[10] Aggarwal S, Johnson S, Saloni D, Hakovirta M. Novel 3D printing filament composite using diatomaceous earth and polylactic acid for materials properties and cost improvement. Composites Part B: Engineering. 2019;**177**

[11] Carll AP et al. Inhalation of printer-emitted particles impairs cardiac conduction, hemodynamics, and autonomic regulation and induces arrhythmia and electrical remodeling in rats. Particle and Fibre Toxicology. 2020;**17**(1):1-22. DOI: 10.1186/s12989-019-0335-z

[12] Galante R, Figueiredo-Pina CG, Serro AP. Additive manufacturing of ceramics for dental applications: A review. Dental Materials. 2019;**35**(6):825-846. DOI: 10.1016/j.dental.2019.02.026

[13] Le Duigou A, Fruleux T, Matsuzaki R, Chabaud G, Ueda M, Castro M. 4D printing of continuous flax-fibre based shape-changing hygromorph biocomposites: Towards sustainable metamaterials. Materials and Design. 2021;**211**:110158. DOI: 10.1016/j.matdes.2021.110158

[14] Momeni F, Sabzpoushan S, Valizadeh R, Morad MR, Liu X, Ni J. Plant leaf-mimetic smart wind turbine blades by 4D printing. Renewable Energy. 2019;**130**:329-351. DOI: 10.1016/j.renene.2018.05.095

[15] Ding Z, Yuan C, Peng X, Wang T, Qi HJ, Dunn ML. Direct 4D printing via active composite materials. Science Advances. 2017;**3**(4):1-6. DOI: 10.1126/sciadv.1602890

[16] Breger JC et al. Self-folding thermo-magnetically responsive soft microgrippers. ACS Applied Materials

and Interfaces. 2015;**7**(5):3398-3405. DOI: 10.1021/am508621s

[17] Gao W et al. The status, challenges, and future of additive manufacturing in engineering. CAD Computer Aided Design. Dec. 2015;**69**:65-89. DOI: 10.1016/j.cad.2015.04.001

[18] Pan Y, Zhao X, Zhou C, Chen Y. Smooth surface fabrication in mask projection based stereolithography. Journal of Manufacturing Processes. 2012;**14**(4):460-470. DOI: 10.1016/j.jmapro.2012.09.003

[19] Yang J, Chen Q, Chen F, Zhang Q, Wang F, Fu Q. Realizing the full nanofiller enhancement in melt-spun fibers of poly (vinylidene fluoride)/carbon nanotube composites. Nanotechnology. 2011;**22**(355707):1-10

[20] Zhang Y, Hao L, Savalani M, Harris R, Tanner K. Characterization and dynamic mechanical analysis of selective laser sintered hydroxyapatite-filled polymeric composites. Journal of Biomedical Material Research Part B Applied Biomaterials. 2008; **86a**:607-616

[21] Espera A Jr, Valino A, Palaganas J, Souza L, Chen Q, Advincula R. 3D printing of a robust Polyamide-12-Carbon black composite via selective laser sintering: Thermal and electrical conductivity. Macromolecules Materials and Engineering. 2019;**182**(1800718):1-8

[22] Burr E. 8 Great Innovations Made Possible Thanks to 3D Printing. ALL3DP; 2019

[23] 3D Printed Titanium Helps Make This Drone Ready to Fight Fires. 3DPRINT.COM; 2015

[24] Airbus. Pioneering Bionic 3D Printing. 2016

[25] Materialise. Airbus Gets on Board with 3D Printing

[26] Madeleine P. Airbus Qualifies Materialise and EOS to Make 3D Printed Flight-Ready Parts. 2021

[27] Goehrke S. Stratasys To 3D Print More Airbus Cabin Components. 2021

[28] Dan S. Airbus had 1,000 Parts 3D Printed to Meet Deadline. BBC;

[29] Altus K. First 3D-Printed House in US Assembled in Hours. *Fox Buisness*; 2022

[30] Vihaan Y. The Cost of 3D Printed Houses in 2021. 3DRIFIC; 2022

[31] Printed Car Parts: 3 Examples of Innovative 3D Printing at Porsche. NextLevelGermanEngineering; 2020

[32] Innovative pistons from a 3D printer for increased power and efficiency. Porsche Newsroom; 2020

[33] Prototype for small-series production: Electric drive housing from a 3D printer. Porsche Newsroom; 2020

[34] Haidiezul AHM, Aiman AF, Bakar B. Surface finish effects using coating method on 3D printing (FDM) parts. IOP Conference Series: Materials Science and Engineering. 2018;**318**:1

[35] Willis KDD, Wilson AD. InfraStructs: Fabricating information inside physical objects for imaging in the terahertz region. ACM Transactions on Graphics. 2013;**32**:4

[36] Chen Y, Ma Z, Zhou H. 3D Print-Scan resilient localized mesh watermarking. In: 2021 IEEE International Workshop on Information Forensics and Security (WIFS). 2021

[37] Gao Y, Wang W, Jin Y. ThermoTag: A hidden ID of 3D printers for

fingerprinting and watermarking. In: IEEE Transactions on Information Forensics and Security. 2021. pp. 2805-2820

[38] Yang XCH, Leow WR, Wang T, Wang J, Yu J, He K, et al. 3D printed photoresponsive Devices based on shape memory composites. Advanced Materials. 2017;**29**:1627

[39] Kotikian A, Truby RL, Boley JW, White TJ, Lewis JA. 3D printing of liquid crystal elastomeric actuators with spatially programed nematic order. Advanced Materials. 2018;**30**:6164

[40] Nadgorny M, Xiao Z, Chen C, Connal LA. Three-dimensional printing of pH-responsive and functional polymers on an affordable desktop printer. ACS Applied Materials Interfaces. 2016;**8**(42):28946-28954

[41] Gladman AS, Matsumoto EA, Nuzzo RG, Mahadevan L, Lewis JA. Biomimetic 4D printing. Nature Materials. 2016;**15**:413-418

[42] Momeni F, Hassani SMM, Liu NX, Ni J. A review of 4D printing. Materials and Design. 2017;**122**:42-79

[43] Momeni F, Jun N. Laws of 4D printing. 2018

[44] Kang HW, Lee SJ, Ko IK, Kengla C, Yoo JJ, Atala A. A 3D bioprinting system to produce human-scale tissue constructs with structural integrity. Nature Biotechnology. 2016;**34**:312-319

[45] 4D Printing for automotive industry applications. Journal of The Institution of Engineer (India): Series D. 2021;**102**:521-529

[46] Ryan KR, Down MP, Banks CE. Future of additive manufacturing: Overview of 4D and 3D printed smart and advanced materials and their applications. Chemical Engineering Journal. 2021;**403**:126162. DOI: 10.1016/j.cej.2020.126162

[47] Tao R, Ji L, Li Y. 4D printed origami metamaterials with tunable compression twist behavior and stress-strain curves. Composites Part B: Engineering. 2020;**201**:108344

[48] Yang C, Boorugu M, et al. 4D printing reconfigurable, deployable and mechanically tunable metamaterials. Materials Horizons. 2019

[49] Yuan L, Pan Z, et al. Fabrication of metallic parts with overhanging structures using the robotic wire arc additive manufacturing. Journal of Manufacturing Processes. 2021;**63**:24-34

Chapter 3

Multiscale Modeling Framework for Defect Generation in Metal Powder Bed Fusion Process to Correlate Process Parameters and Structural Properties

Suchana Akter Jahan and Hazim El-Mounayri

Abstract

Powder Bed Fusion (PBF) is one of the most popular additive manufacturing methods employed extensively to fabricate complex parts especially in industries with stringent standard criteria, including aerospace, medical, and defense. DMLS/PBF fabrication of parts that is free of defects represents major challenges. A comprehensive study of thermal defects, contributing parameters, and their correlation is necessary to better understand how process specifications initiate these defects. Monitoring & controlling temperature and its distribution throughout a layer under fabrication is an effective and efficient proxy to controlling process thermal evolution, which is a completely experimental technique. This being highly costly specifically for metal printing, computer-based numerical simulation can significantly help the identification of temperature distribution during the printing process. In this paper, a multiscale modeling technique is demonstrated with commercially available software tools to correlate the defect generation in metal PBF process and significant process parameters. This technique can help efficiently design the process setting in addition to or even absence of experimental monitoring data. This research work is a part of a larger project of closed-loop control strategy development using physics-based modeling and graph-based artificial neural network implementation for reducing thermally induced part defects in metal 3D printed process.

Keywords: powder bed fusion, process parameters, defect generation, thermal anomalies, artificial neural network, in-situ monitoring, feedback control

1. Introduction

Additive Manufacturing (AM), also known as Rapid Prototyping and 3D Printing is a three-dimensional fabrication process, executed by adding materials in layers. This is a revolutionary product development and manufacturing method, especially in the age when we are experiencing a mew industrial revolution. Small, relatively simple products may only make use of AM for visualization models, while larger, more complex products with greater engineering content may involve AM during numerous stages and iterations throughout the development process [1]. Among many different classes of AM processes, Direct Metal Laser Sintering (DMLS) is a widely used metal part manufacturing method. This process is carried on by using laser power to melt powder metal particles leading to a complete print of the part using the desired 3D solid CAD model data by a layer-by-layer process [2]. The key is to melt the material in a controlled fashion without creating a high accumulation of heat, so that when the energy source is removed, the molten material rapidly solidifies again. Recent involvement of large companies in developing metal AM processes have opened up the market significantly. As a result, machine accuracy, speed, cost, and quality of production have become apparent crucial factors in metal 3D printing nowadays. High quality metal parts with complex geometry can be produced in PBF process. At the same time the manufacturer needs to ensure part-quality, consistency, and competitive pricing to run and sustain a successful metal printing business. Here comes the requirement for deeper understanding of the process so that each and every step can be improved, optimized and controlled for higher efficiency, reliability, and production quality. The metal printing technology is in practice for more than a decade now, yet researchers are working till date to understand the physics of powder bed fusion process, and this is an ongoing pursuit. In this paper, a multiscale modeling technique is described which aims at correlating the PBF process parameter and their impact on single layer as well as subsequent printing defects on build part.

2. Background

Powder bed Fusion (PBF) or Direct metal laser sintering (DMLS) is an additive manufacturing process to build fully functional rapid metal prototypes and production parts. A selective sintering process is carried on using laser power to melt powder metal particles leading to a complete print of the part by a layer-by-layer process [1]. DMLS can be used to create parts from a variety of metal alloys. This makes it more popular than many other 3D printing techniques that are designed to work with specific metal alloys or polymer-based materials. One big advantage of DMLS is that custom manufactured parts made using this process tend to be free of internal defects and residual stress that typically occur in parts that have been made using more traditional manufacturing methods. This ability to create defect-free parts is critical when parts are to be used in a high-stress environment such as the automotive or aerospace industries.

2.1 Defects in PBF build parts

Several common defects are encountered in metal PBF processes, which lead to weakening of mechanical properties of the build part. These defects occur during fabrication and/or post -processing operation. The presence of defects limits the industry-wide spread of this technology as a result of insufficient repeatability, reliability, and precision. Such defects can be classified and analyzed based on how they affect the printed part, how they generate, how common and significant they are for the overall quality of the build part etc. In addition to that, different process parameters influence the generation and propagation of different anomalies in the print. Many studies have been conducted to identify the defects in AM processes [3–5].

In a recent work [6], the common defects in additive manufacturing have been classified and reviewed on the basis of geometry, surface quality, microstructure, and mechanical properties. For example, on the basis of geometry and dimension, there can be (a) geometry inaccuracy (form deviation) and (b) dimension inaccuracy (size deviation). Again, common surface quality related defects are surface roughness, balling, surface oxidation etc., while anisotropy, heterogeneity, porosity etc. are microstructure related defects found in 3D printed builds. A comprehensive classification is shown in **Figure 1** [6].

2.2 How do process parameters affect printing defects

The significant process parameters in the metal powder bed fusion (also known as selective laser sintering process) such as laser power, scanning speed, hatching distance, scanning strategy etc. affect the generation of printing defects. It is crucial to identify the combination of these parameters to obtain the required level of quality of the product. For instance, higher laser power increases the thermal shrinkage, and higher scanning speed hatch spacing lowers the thermal shrinkage [7]. The laser power and scanning strategy contributed to the temperature variation, that leads to non-uniform shrinkage in a particular layer [8]. Part weight, build chamber temperature, cooling rate, layer thickness and material can affect thermal shrinkage in a way that shrinkage decreases with increasing layer thickness, part bed temperature, and interval time [9–11].

Surface roughness depends on numerous parameters, some of which can be controlled, while few others are not controllable from the designer and operator's perspective. For lower scan speeds, the average roughness decreases with increase in speed, while in the higher speed range, it remains constant [11–16]. Moreover, warping and distortion that impact the surface quality, are highly dependent on thermal phenomena during the printing process. In addition to laser power and scanning speed, the thermal gradient between scanning zones can impact the quality of each fabricated layer. Surface geometry and fundamental geometric features of orientation, thickness etc. also have impact of geometric errors and surface quality [17]. Smaller hatch spacing seems to be beneficial in this context as it induces gradual temperature increase of powder bed and slower cooling rate. In addition to this, the initial powder spreading in LPBF also influences the layer quality, and thus consequentially impacts the porosity of produced parts [18].

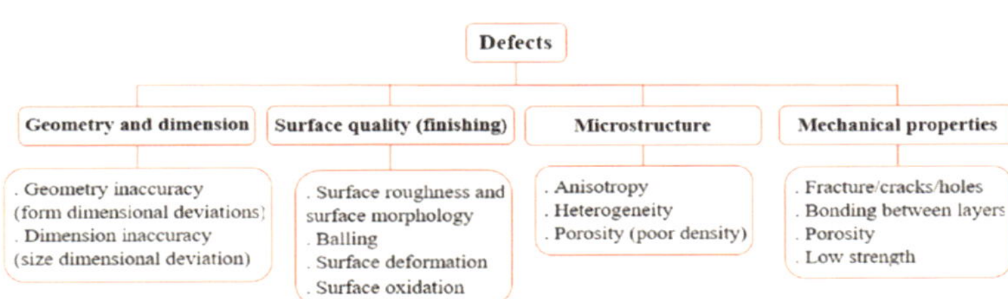

Figure 1.
Defects in metal PBF.

Typical manufactured components using traditional manufacturing methods (milling, drilling, surface grinding etc.) comes from a solid building block of material and they are not porous unless porosity is induced by design. But in additive manufacturing, porosity of the build parts is a very common occurrence. Poor wetting, powder packing density, gas flow condition, entrapped gas etc. play vital roles behind the unwanted porosity of printed parts.

Among the numerous process parameters that contribute to defect generation, most vital ones are laser power, laser scanning speed, laser spot size, powder size, layer thickness, external pressure and material's absorptivity. It is important to understand how the process parameters impact the generation of defects and what are the signatures that relate to the defects. The process-structure–property (PSP) relationships have been under discussion and research works are published earlier as well [19–21]. In recent years, use of machine learning and smart manufacturing has started to be used in AM field [22–31], which also provides inspiration to use that for our research goal.

3. Generation of defects

The two major phenomena that contributes to the generation of defects in powder bed fusion are **Keyhole** and **Spattering**. In this section, these modes are discussed to understand the underlying physics of why and how a defect may occur in a layer or whole printed part. In a typical laser processing case, two modes of heat transfer participate in defect generation: namely, conduction mode and keyhole mode [32]. Conduction is melting is controlled by heat conduction. In keyhole mode, input power density is sufficient or high enough to vaporize the metal powder. It then creates a much deeper hole or cavity compared to conduction mode. Collapse of such cavities can leave voids in printed parts [33], hence, conduction mode of heat transfer is desired in laser additive manufacturing.

3.1 Keyhole mode

As mentioned earlier, keyhole is a major cause of forming pores and voids. Basic criterion of identifying the keyhole mode is shown in Eq. (1)

$$\frac{2\mathrm{d}}{\mathrm{w}} \geq 1 \tag{1}$$

where d is the remelted depth, w is the melt pool width, and the quantity, d/w, is the normalized depth [34–36]. There are few other empirical formulae and methods proposed by researchers [37]. These enable the designers to identify the preferred conduction mode in L-PBF conveniently. By experimental methods, it was found that normalized depth is proportionally related to product of power density and square root of laser interaction time [38]. It obeys Eq. (2) as following:

$$\frac{d}{w} \propto \frac{P}{\pi\sigma^2} \times \sqrt{\frac{2\sigma}{v}} \tag{2}$$

where P is laser power, σ is laser spot size, and ν is laser scan speed.

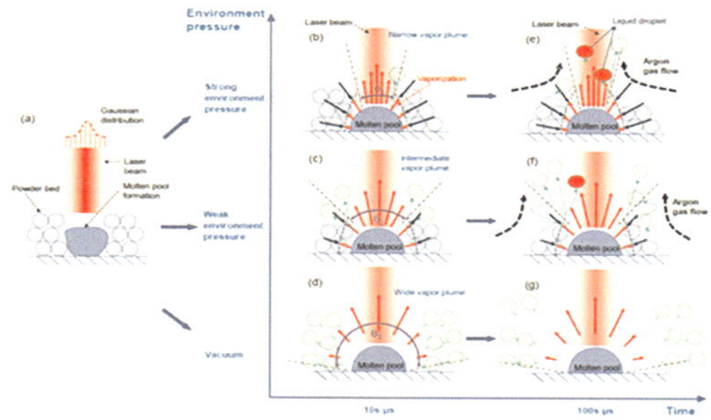

Figure 2.
Pressure and time dependent spattering mechanism.

3.2 Spattering phenomenon

Spattering is a physical phenomenon in laser powder bed fusion process observed from experiments [39, 40]. It is considered to be the major cause of the structural defects in the printed products. This is a complex physical phenomenon that requires experimental procedures to observe and challenging to model properly in a FEA computer numerical analysis.

From previous experimental tests, it is found that some spatters have a propensity to merge together and form larger particles. Through their comparison study between multi-laser and single laser scanning, Andani et al. [41] found that high number of working lasers can induce higher recoil pressure above melting pool and as a result more spatters are ejected for molten pool. In case of stationary laser impulse, as time goes by, vaporization occurs after melting and generates intensive vapor jet that ejects metal powders. With the surrounding pressure of inert gas decreasing, it forces the particles surrounding the molten pool to move forward. In this manner, metal powders are ejected with a large divergence angle as vapor can expand freely. This physical event can be observed as "Spattering" by using high-speed X-ray monitoring. **Figure 2** [42] shows a schematic of pressure and time dependent spattering mechanism.

4. Multiscale modeling technique

As mentioned earlier, spattering mechanism is difficult to model using traditional simulation approach due to its multi-phase nature and complexity in physics. This applies to the overall additive manufacturing or metal printing process as well. Molecular Dynamics (MD) can be an option in this area, considering only the atomic motion without any other information such as thermal conductivity or specific heat of the material. But, due to its nanoscopic nature, available computational capacity of today falls way behind the complexity of any feasible MD model. Some experimental techniques can be used as demonstrate in [43] to understand he metallurgical defects in PBF parts. But investigate generation of defects due to the process itself is significantly difficult. Computational Fluid Dynamics (CFD) modeling can be applied to simulate sintering process involved in L-PBF, as this can demonstrate the molten pool dynamics. But the fact that solid metal powders can only be stationary in CFD models, can limit the scope of complete simulation of AM process to understand the process-structure–property relationships. On the other hand, a simple meso-scale simulation using finite element analysis can help to correlate the build part defects (such as, deformation, buckling, holes, staircase effect etc.) with AM process parameters. This in turn initiates the idea of a multi-scale modeling scheme for overall understanding of the AM process.

A CFD modeling tool can correlate the process parameters (laser power, scanner speed etc.) to the defect generation mode or phenomenon (e.g., keyhole and spattering). Finite element simulation can predict temperature distribution over each layer with appropriate beam size and diameter, using material properties of the metal powder and subsequent build. The complexity of simulation depends on the available physics on the modeling scheme. However, these computational studies and any other experimental observations are not sufficient enough to create a comprehensive design and optimization method. To obtain a generalized process parameter optimization technique, a surrogate model is desirable to alleviate the overbearing requirements of frequent experiments/ simulations.

An alternative solution of this impasse is application of machine learning (ML) model. A supervised machine learning algorithm would be able to create an artificial neural network intelligent enough to predict the defect generation and hence recommend suitable combination of process parameters to generate flawless/ near- flawless printed parts. But performance of such algorithm depends on a well training program with appropriate input–output data. This data would come from numerical simulation and/or experimental testing. The multiscale modeling with different level of simulation will add to the experimental data that can be obtained in a real-time laser powder bed setting.

This multiscale modeling work is divided into three major steps:

- Micro-scale Computational Fluid Dynamics Analysis (additive metal powder size varies in the range of 50–100 μm. The simulation covers powder bed preparation, deposition and spreading and melt pool analysis in micro-scale).

- Meso scale Finite Element Analysis (focuses on macroscale geometry, stress, strain, deformation and build time).

- Validation study.

4.1 CFD analysis

A commercially available CFD software named ***FLOW-3D*** specializing in 3D transient flows with free surfaces is used in this study. It follows Volume of Fluid (VoF) method and TruVOF algorithm. The physics behind general welding and laser melting in PBF is similar, hence it is used in the CFD modeling scheme. Key factors involved here are laser beam motion, shield gas pressure, laser heat flux profile distribution, evaporation pressure and multiple laser reflections.

With the input of material properties of metal powder, powder size, bed size, the first step of the simulation is powder bed preparation. After the bed spreading is simulated, we can input the process parameters such as laser power, beam diameter etc. for the designed geometry and it will subsequently complete the laser melting simulation. **Figure 3** the laser melting of metal and subsequent melt pool formation with temperature distribution [44]. Keyhole-induced porosity formation is observed in **Figure 4** [45].

Understanding the evolution of melt pool depending on varying process parameters can provide information on temperature distribution and porosity formation as shown in **Figure 5** [44]. This is a crucial information for the multiscale-surrogate model. Rise in recoil pressure at the bottom of keyhole and increased surface tension

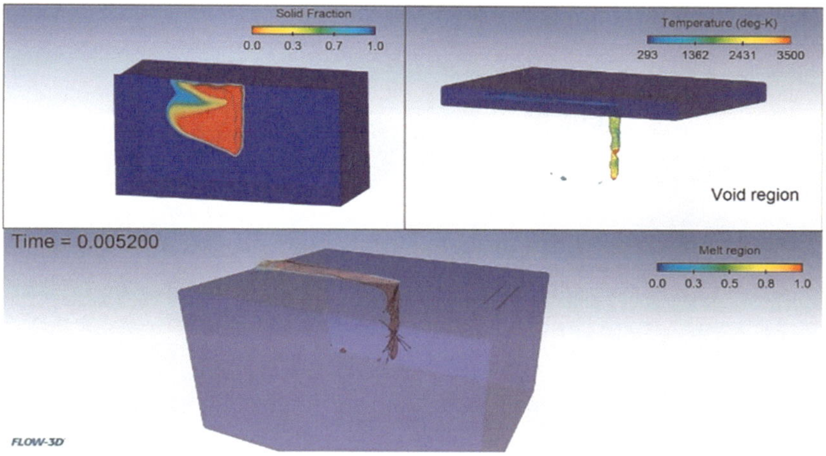

Figure 3.
Laser keyhole welding modeling using CFD.

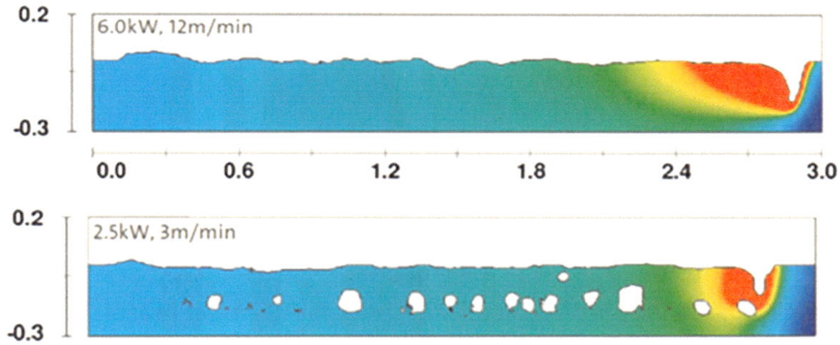

Figure 4.
Keyhole induced porosity formation in L-PBF process.

at the upper region generates an irregular pore. The pore is then pushed to the back of melt pool by string downward flow and then it gets trapped by the advancing solidification front. Using this aforementioned CFD model, we can accurately represent the fluid flow at the melt pool at 1–10 μm length scale [44]. It also demonstrates defective design space and melt pool geometries, predicts compositional segregation and phase nucleation and growth.

4.2 FEA model

A recently developed and launched suite in ANSYS workbench called "ANSYS Additive" provides capabilities to simulate the complete metal additive printing process in L-PBF method. In addition to the common uses of creating and optimizing design solutions for AM purposes, it can also be used for understanding the metallurgical properties of the printed parts with porosity and microstructure prediction. Accessible data throughout the process ensures the traceability and hence enhances the feasibility of a parametric study using DOE (Design of experiments) technique.

In this method, we incorporate the layer-wise structure development and time discretization (exposure time of single layers is in milliseconds, but total build time

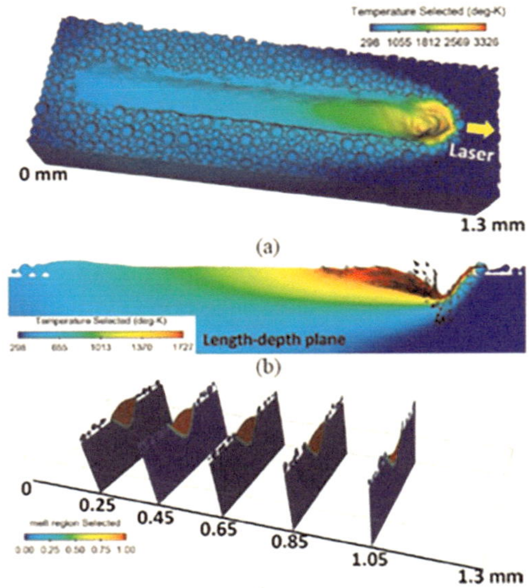

Figure 5.
Melt pool formation in L-PBF, (a) 3D view, (b) 2D view, (c) sectional view.

can be hours) by using a detailed model for single layer and global model for whole structure (**Figure 6**). This is called lumped layer approach. The details of the theoretic development of this model can be found in ANSYS Additive training website [46].

Necessary material information to perform this simulation are density, thermal conductivity, heat capacity, structural properties such as Young's modulus, Poisson's ratios, thermal expansion co-efficient, and stress–strain data. Material properties of common AM metals can be obtained from pre-existing database of Workbench Additive and/or modified with specific case study or experimentally driven data.

Using this technique, it is possible to simulate the building of whole structure, without the complexity of looking into the details of each layer formation and related void generations. As we already know the parametric relationships between defect generation and process parameters from the aforesaid CFD model in Section 5.1, this FEA model complements it with the whole structural information of the printing process.

The main challenge for the FEA simulation is local discretization, by which we mean the dimensions of the laser spot are in μm, whereas the dimension of the whole structure is in cm. Moreover, exposure to a single layer occurs in ms (milli second), but the full built takes longer time, even couple of hours. This issue can be taken care of by applying a lumped layer approach in ANSYS additive. A quick simulation catches global stress/strain as well as the distortion that takes place during printing. The whole structure is simulated in a global simulation model, where several subsequent simulations are done with status and boundary conditions are updated consequentially. For such simulations, a few assumptions are made. These are:

- thermal and structural physics are uncoupled,
- no use of laser beam (consider whole layer at a time),

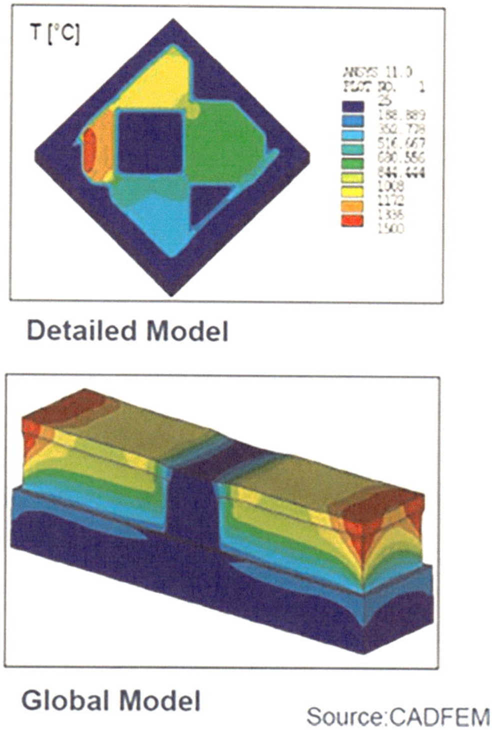

Figure 6.
Lumped layer approach; top: Single layer detailed model, bottom: Whole structure global model.

- several powder layers are lumped into one FE layer, and
- no melted powder is modeled here [46].

This is acceptable as we are focusing on the melt pool in the CFD simulation, and the combined modeling scheme will provide a comprehensive identification of all the relevant phenomena in the printing process.

A sequentially weakly coupled transient thermal-structural analysis is performed in this FEA model. Similar to conventional structural or thermal analysis in ANSYS workbench, we can import any kind of CAD file to this additive suite. Next steps include body cartesian mesh generation, creating named selections for build, support, and base. The simulation wizard is capable of automatically generating the support and base, so the designer only needs to consider the design of build part. After that, the build settings should be defined. This includes machine setting, deposition thickness, hatch spacing, laser speed, time between layers etc. Thermal boundary conditions include preheat temperature, gas/powder temperature, convection coefficients, and cooldown temperature (usually room temperature). Common deposit thickness is 0.001 mm–0.1 mm. A very large model will need higher computation power and time, in such cases, High performance computing (HPC) will help.

Using result tracker for temperature and displacement, the user can control the progress of print process during the solution as shown in **Figure 7** [46]. Moreover, it is possible to switch between automatic and manual mode for result tracker.

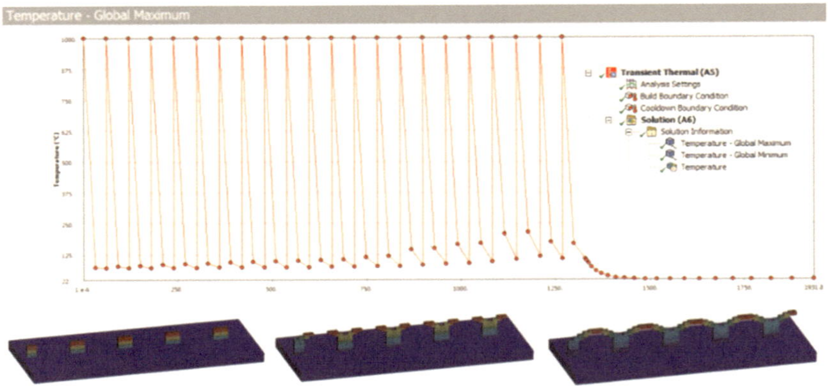

Figure 7.
Result tracking during solution in FEA model, top: Tracking global maximum temperature, bottom: Build progression shown from left to right.

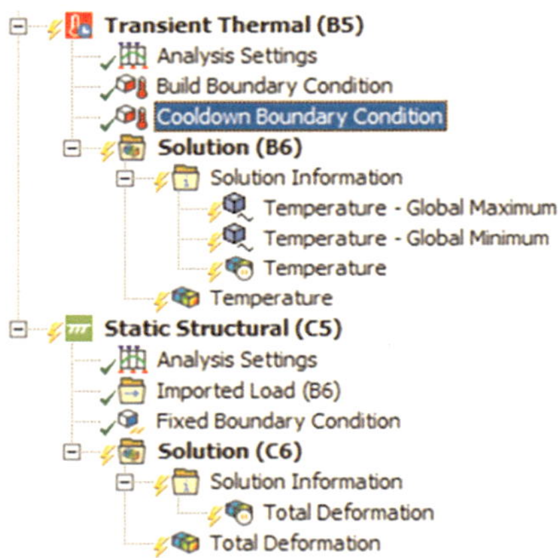

Figure 8.
Typical model tree of coupled AM FEA study.

As mentioned earlier it is a weakly coupled FEA analysis, a transient thermal analysis is first conducted with appropriate properties and boundary conditions. Then the results are fed into a structural analysis model, and we would finally obtain the deformation, bucking etc. on the whole structure. A general model tree is shown in **Figure 8** for easy understanding of reader.

4.3 Validation

To develop a reliable simulation model and build a dependable training data set, it is necessary to validate the model and its input–output information. This requires experimental testing of the designed case studies in a real-time metal additive printing setup. There are several small-scale additive metal printers are available in the market which are cost-effective, but ideally, they do not reciprocate all the complexities of a full-size metal printer. Hence, we will perform experimental testing on a real time metal PBF setup to validate the multiscale numerical model.

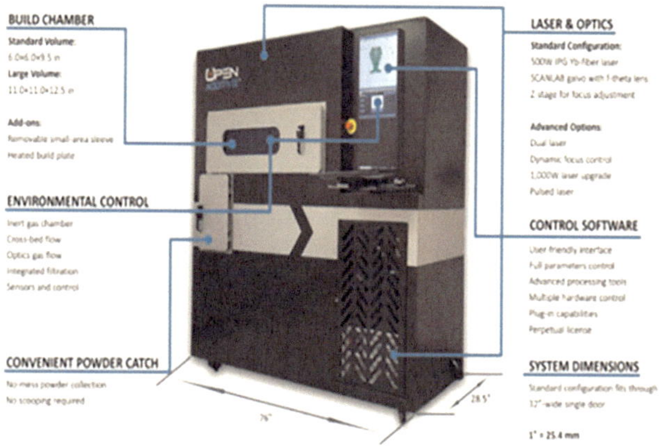

Figure 9.
Open architecture 3D metal printer PANDA.

These experiments will serve two purposes:

- Provide thermography data for building a training data set.
- Validate simulation model by replacing few user cases with specified process parameter combinations.

Experiments will be conducted in an open-structure metal AM printer, PANDA 11 (**Figure 9**). This open structure allows the system to monitor the build system and track temperature data for each layer. This system also enables closed-loop feedback operated online monitoring and control system. This is a part of the ongoing research that the authors are working currently. Details of the project can be found in recent publication of this research group [1]. This project involves understanding the physics of defect generation in metal powder bed fusion and using machine learning (ML) algorithm to implement the knowledge in automated process parameter selection to minimize printing defects. The machine learning algorithm under consideration is a graph based spatio-temporal convolutional neural network (ST-GCN) that will be trained using the results obtained in CFD and FEA modeling. The code will also incorporate genetic algorithm and/or game theoretic model to optimize the process parameters in order to minimize the defect generation. Once the ML code is trained and tested, it will be implemented using the device driver of the Open additive machine (shown in **Figure 9**). During operation, an online monitoring system using IR camera will be used to track the thermal history. This spatio-temporal temperature data will work as input to the ML algorithm, and finally using optimization theory, the device driver will receive information on optimized combination of process parameters, that will be activated for the next layer of printing. This is a novel idea for controlling the defect generation in metal additive manufacturing using process parameters and physics-based understanding of the process.

5. Data collection

The information obtained from FEA study and validation study can be used in order to design and develop a spatio-temporal neural network. This artificial intelligence algorithm can provide optimized process parameters to produce parts with minimum defects. The experimental procedure being costly and designed to be at the end of the project, we started with initial FEA study at CFD level. We have created a Design of Experiment (DOE) study using CFD analysis. We have used two different material powder, three different laser power setting and three different scanning speed settings to perform the DOE. Moreover, we are considering single pass single layer melting only. The main purpose of this DOE is to investigate the diverse data obtainable from the simulations and understand impact of process parameter on the temperature map on PBF processes. So, there are 2 x 3 x 3 = 18 cases in design of experiments using full factorial method. **Table 1** shows the DOE setup.

In addition to the defined input variables for the case study, i.e., material type, laser power and speed, many other parameters are also needed to setup the CFD simulation in ***FLOW-3D AM®***. The simulation setup is quite critical as the melting process itself is very complex. Basic setup steps are described here. Interested readers can visit here to obtain more in-depth knowledge and tutorials to use for individual studies.

- In ***FLOW-3D AM®*** start a melting simulation in workspace, using CGS unit system and kelvin for temperature.

- In global settings, set finish time for simulation. We have used 0.0005 s for the initial simulations. But it can be varied as per desired setup.

- Bubble and phase change physics activated with constant pressure of vaporization where

    - Pressure = 1.0e6 dyne/cm^2,

    - Heat transfer co-efficient =1.0e5 erg/cm^2/s/K

    - Gamma = 1.4.

- Density is set as function of temperature.

Case number	Material	Laser power (W)	Scanning speed (cm/s)	Case number	Material	Laser power (W)	Scanning speed (cm/s)
1.1	Inconel718	280	220	2.1	PH-316 SS	280	220
1.2	Inconel718	280	300	2.2	PH-316 SS	280	300
1.3	Inconel718	280	350	2.3	PH-316 SS	280	350
1.4	Inconel718	400	220	2.4	PH-316 SS	400	220
1.5	Inconel718	400	300	2.5	PH-316 SS	400	300
1.6	Inconel718	400	350	2.6	PH-316 SS	400	350
1.7	Inconel718	750	220	2.7	PH-316 SS	750	220
1.8	Inconel718	750	300	2.8	PH-316 SS	750	300
1.9	Inconel718	750	350	2.9	PH-316 SS	750	350

Table 1.
DOE case setup for CFD simulations.

- Gravity, in Z direction, −981 cm/s^2.
- Surface tension, laminar viscous flow and solidification mode activated.
- Surrounding fluid is air at 15°C.
- Meshing size of cell 0.0005 cm.
- The simulation area can be created using particle size, density, maximum number of particles, packing density etc. using FLOW 3D and then converted into STL into particle to STL converter. This STL file can be used as base for all case studies as "particle bed". Additional fluid region can also be incorporated.
- Laser power, velocity, lens shape and size can be defined using ***FLOW 3D WELD*** module. In this study we have used circular lens with focal distance

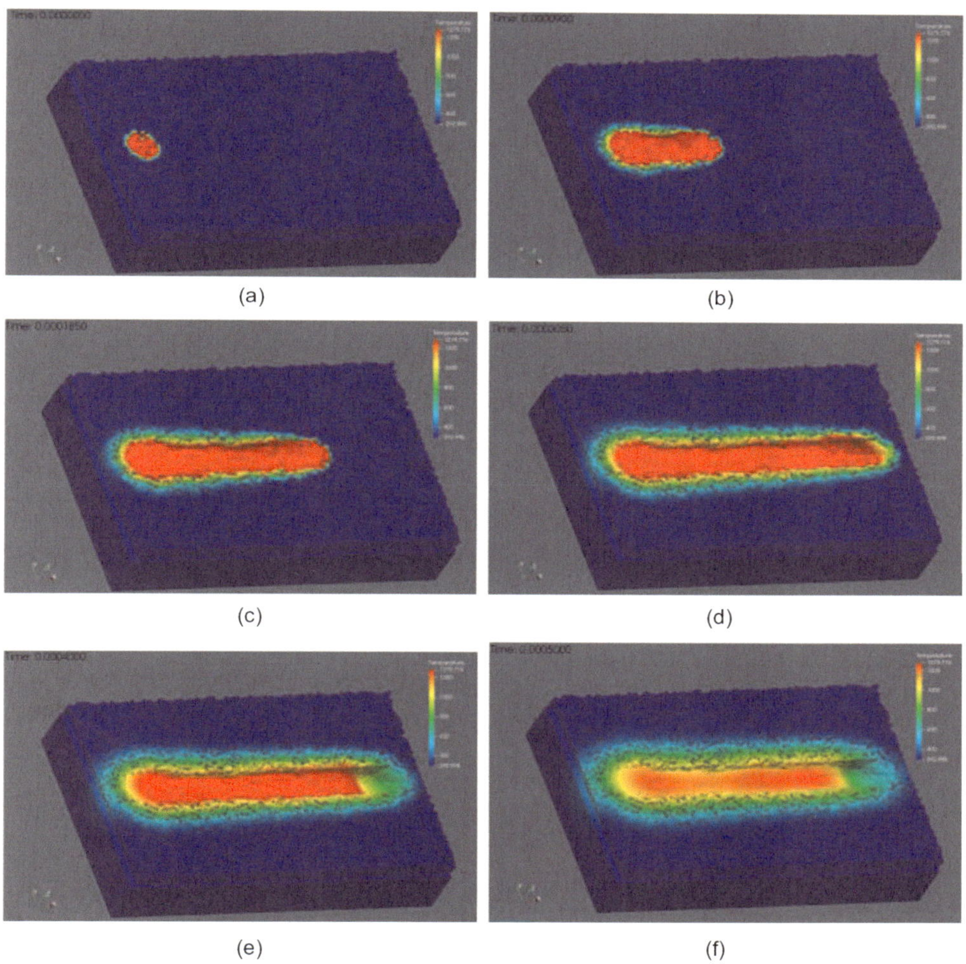

Figure 10.
Laser progression and temperature profile map (a) to (f) shows progression in time domain.

0.8 cm, radius of 0.01 mm and spot radius of 0.003 cm. Only X directional velocities are used, they are zero in Y and Z directions.

- Materials are used only Inconel-718 and PH-316 Stainless Steel in this case study, and the built-in property directory is utilized as well. In specific studies, we can also edit any relevant properties of material as needed.

6. Results and discussion

The results obtained from FLOW 3D AM can be analyzed using simple rendering in the program itself or can be further post processed using their FLOW 3D POST

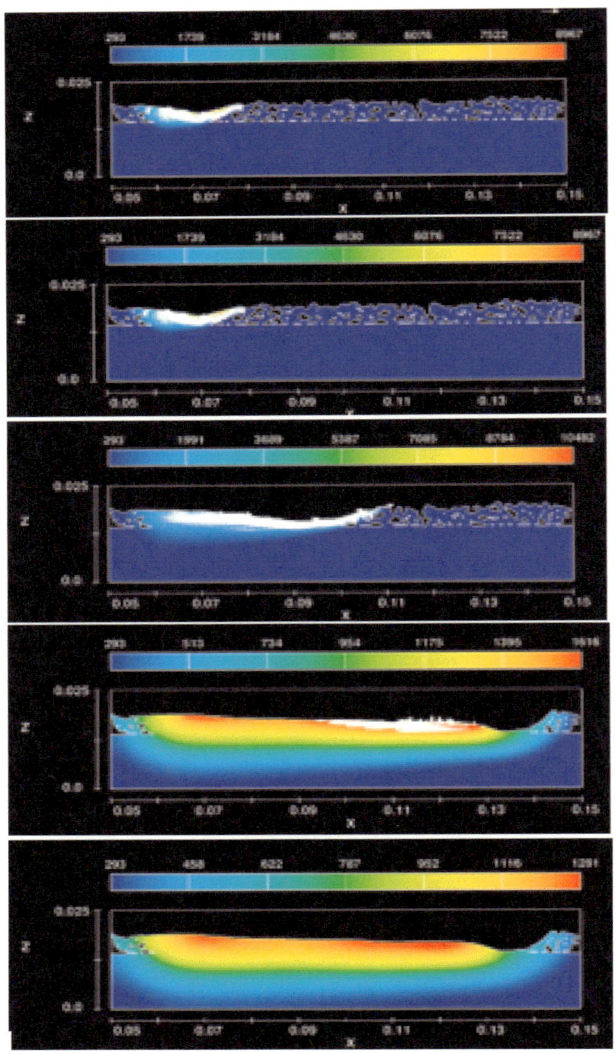

Figure 11.
Representative melt pool on 2D plane (progression from top to bottom).

package. It is necessary to include/activate the required outputs whilst setting up the simulation, so that it is easier to post process. After completion of simulations, usually the flsgrf.melting files are rendered and used in further post processing. As we have several cases in the DOE, we are only discussing a few sample cases in this paper due to limitation of space.

In **Figure 10**, we are attempting to provide a visualization of how the laser moves starting at the left side and the temperature distribution changes with time. We are including only 6 sample timesteps to demonstrate the progress. In **Figure 10(a)**, we capture the time step at t = 0.000005 s. The progression occurs through **Figure 10(d)**, at t = 0.000305 s, where the laser has reached the end with melt pool at highest temperature. Moving further in time, on **Figure 10(e)** at t = 0.0004 s, we can see the melt pool gradually is cooling off. This is clearly visible at the right.

Next, in **Figure 11**, the 2D cross section (X-Z plane) is presented on progressive time step, from top to bottom. Only few timestep captures are only provided here for ease of conception. Important feature to notice here is in such a 2D view, we can easily identify the voids or porous structures that are left behind during the melting process. With this knowledge we can further investigate to correlate these to defect generation and thus optimize process parameters accordingly. In this technique, studying the microscopic density, energy density etc. will also contribute profoundly and these spatio-temporal data are also available.

7. Future work

The work described in this paper is a part of research project that aims at developing an intelligent closed loop feedback-controlled 3D metal printing system. The simulation framework will be used to generate training data for neural net training in addition to real life experimental results. With sufficient training data, the developed neural network can serve as a significant tool to minimize anomalies in metal additive manufacturing. These will be the foundation of the automated printing system that will choose the process parameters in order to minimize defects and deformities. This project will be a new approach in metal printing industry that will ultimately improve the quality and quantity of product.

8. Conclusion

Additive manufacturing has been in practice for decades, yet continuous developments are being carried out due to its complex nature. In this paper, a modeling framework is presented to help understand the physical phenomenon behind the printing technique and correlate the process parameters with generated defects. The framework includes the computer numerical modeling techniques to be used to simulate the macro and micro scale phenomenon of powder bed fusion process. This is designed in a way to identify temperature signatures and anomaly development occurrences to help identify the root causes of defect generation and in a way categorize the optimal process parameters. The next step includes experimental validation of the techniques. This paper contains relevant information on the experimental testing on industrial printing bed. Being in the initial stage of the project, though we have not completed the experimental studies, and we believe any miniature scale experimentation would not do justice to the presented framework. We expect the context provided in this paper will help research and industry community to improve the state of the art in this field. Industry needs minimum defects and maximum quality with sufficient reliability to be able to flourish in international market. The authors hope this paper would help metal additive manufacturing industry in this matter.

Author details

Suchana Akter Jahan* and Hazim El-Mounayri
Purdue School of Engineering and Technology, Indiana University Purdue University Indianapolis, Indianapolis, IN, USA

*Address all correspondence to: sjahan@iu.edu

References

[1] Jahan S, El-Mounayri H. Game theoretic modelling approach for optimizing direct metal laser sintering process parameters using artificial neural network. In: ASME International Mechanical Engineering Congress and Exposition. Vol. 84492. American Society of Mechanical Engineers; Virtual Conference, 2020. p. V02BT02A024

[2] Singh DD, Mahender T, Reddy AR. Powder bed fusion process: A brief review. Materials Today: Proceedings. 2021;**46**:350-355

[3] Brennan M, Keist J, Palmer T. Defects in metal additive manufacturing processes. Journal of Materials Engineering and Performance. 2021; **30**(7):4808-4818

[4] Li R, Jin M, Paquit VC. Geometrical defect detection for additive manufacturing with machine learning models. Materials & Design. 2021;**206**: 109726

[5] Chen Y, Peng X, Kong L, Dong G, Remani A, Leach R. Defect inspection technologies for additive manufacturing. International Journal of Extreme Manufacturing. 2021;**3**(2):022002

[6] Malekipour E, El-Mounayri H. Common defects and contributing parameters in powder bed fusion AM process and their classification for online monitoring and control: A review. The International Journal of Advanced Manufacturing Technology. 2018;**95**(1): 527-550

[7] Paul R. Modeling and Optimization of Powder Based Additive Manufacturing (AM) Processes. Cincinnati, USA: University of Cincinnati; 2013

[8] Venuvinod PK, Ma W. Rapid Prototyping: Laser-Based and Other Technologies. Springer Science & Business Media; 2013

[9] Wang X. Calibration of shrinkage and beam offset in SLS process. Rapid Prototyping Journal. 1999;**5**:129-133

[10] Shi Y, Li Z, Sun H, Huang S, Zeng F. Effect of the properties of the polymer materials on the quality of selective laser sintering parts. Proceedings of the Institution of Mechanical Engineers, Part L: Journal of Materials: Design and Applications. 2004;**218**(3):247-252

[11] Wang R-J, Wang L, Zhao L, Liu Z. Influence of process parameters on part shrinkage in SLS. The International Journal of Advanced Manufacturing Technology. 2007;**33**(5–6):498-504

[12] Hauser C. Selective Laser Sintering of a Stainless Steel Powder. Leeds, England: University of Leeds; 2003

[13] Kumar S. Selective laser sintering: A qualitative and objective approach. JOM. 2003;**55**(10):43-47

[14] Hopkinson N, Sercombe TB. Process repeatability and sources of error in indirect SLS of aluminium. Rapid Prototyping Journal. 2008;**14**:108-113

[15] Raghunath N, Pandey PM. Improving accuracy through shrinkage modelling by using Taguchi method in selective laser sintering. International Journal of Machine Tools and Manufacture. 2007;**47**(6):985-995

[16] Ning Y, Wong Y, Fuh J. Effect and control of hatch length on material properties in the direct metal laser sintering process. Proceedings of the Institution of Mechanical Engineers, Part B: Journal of Engineering Manufacture. 2005;**219**(1):15-25

[17] Jones A, Leary M, Bateman S, Easton M. Effect of surface geometry on laser powder bed fusion defects. Journal of Materials Processing Technology. 2021;**296**:117179

[18] Sehhat MH, Mahdianikhotbesara A. Powder spreading in laser-powder bed fusion process. Granular Matter. 2021; **23**(4):1-18

[19] Zhang Y et al. Additive manufacturing of metallic materials: A review. Journal of Materials Engineering and Performance. 2018; **27**(1):1-13

[20] Sames WJ, List F, Pannala S, Dehoff RR, Babu SS. The metallurgy and processing science of metal additive manufacturing. International Materials Reviews. 2016; **61**(5):315-360

[21] Frazier WE. Metal additive manufacturing: A review. Journal of Materials Engineering and Performance. 2014;**23**(6):1917-1928

[22] Majeed A et al. A big data-driven framework for sustainable and smart additive manufacturing. Robotics and Computer-Integrated Manufacturing. 2021;**67**:102026

[23] Wang C, Tan X, Tor S, Lim C. Machine learning in additive manufacturing: State-of-the-art and perspectives. Additive Manufacturing. 2020;**36**:101538

[24] Meng L et al. Machine learning in additive manufacturing: A review. JOM. 2020;**72**(6):2363-2377

[25] Jiang J, Xiong Y, Zhang Z, Rosen DW. Machine learning integrated design for additive manufacturing. Journal of Intelligent Manufacturing. 2020;**33**:1-14

[26] Sing S, Kuo C, Shih C, Ho C, Chua C. Perspectives of using machine learning in laser powder bed fusion for metal additive manufacturing. Virtual and Physical Prototyping. 2021;**16**(3): 372-386

[27] Grierson D, Rennie AE, Quayle SD. Machine learning for additive manufacturing. Encyclopedia. 2021;**1**(3): 576-588

[28] Zhu Q, Liu Z, Yan J. Machine learning for metal additive manufacturing: Predicting temperature and melt pool fluid dynamics using physics-informed neural networks. Computational Mechanics. 2021;**67**(2): 619-635

[29] Liu S, Stebner AP, Kappes BB, Zhang X. Machine learning for knowledge transfer across multiple metals additive manufacturing printers. Additive Manufacturing. 2021;**39**:101877

[30] Zhan Z, Li H. Machine learning based fatigue life prediction with effects of additive manufacturing process parameters for printed SS 316L. International Journal of Fatigue. 2021; **142**:105941

[31] Parmar H, Khan T, Tucci F, Umer R, Carlone P. Advanced robotics and additive manufacturing of composites: Towards a new era in industry 4.0. Materials and Manufacturing Processes. 2021:1-35

[32] Meng L. Machine Learning and Probabilistic Design Framework for Laser Powder Bed Fusion Process. IN, USA: Purdue University Graduate School; 2020

[33] Madison JD, Aagesen LK. Quantitative characterization of porosity in laser welds of stainless steel. Scripta Materialia. 2012;**67**(9):783-786

[34] Eagar T, Tsai N. Temperature fields produced by traveling distributed heat sources. Welding Journal. 1983;**62**(12): 346-355

[35] Matilainen V-P, Piili H, Salminen A, Nyrhilä O. Preliminary investigation of keyhole phenomena during single layer fabrication in laser additive manufacturing of stainless steel. Physics Procedia. 2015;**78**:377-387

[36] Zhang Y. Multi-Scale Multi-Physics Modeling of Laser Powder Bed Fusion Process of Metallic Materials with Experiment Validation. IN, USA: Purdue University; 2018

[37] King WE et al. Observation of keyhole-mode laser melting in laser powder-bed fusion additive manufacturing. Journal of Materials Processing Technology. 2014;**214**(12): 2915-2925

[38] Hann D, Iammi J, Folkes J. A simple methodology for predicting laser-weld properties from material and laser parameters. Journal of Physics D: Applied Physics. 2011;**44**(44):445401

[39] Criales LE, Arısoy YM, Lane B, Moylan S, Donmez A, Özel T. Laser powder bed fusion of nickel alloy 625: Experimental investigations of effects of process parameters on melt pool size and shape with spatter analysis. International Journal of Machine Tools and Manufacture. 2017;**121**:22-36

[40] Simonelli M et al. A study on the laser spatter and the oxidation reactions during selective laser melting of 316L stainless steel, Al-Si10-Mg, and Ti-6Al-4V. Metallurgical and Materials Transactions A. 2015;**46**(9):3842-3851

[41] Andani MT, Dehghani R, Karamooz-Ravari MR, Mirzaeifar R, Ni J. Spatter formation in selective laser melting process using multi-laser technology. Materials & Design. 2017;**131**:460-469

[42] Guo Q et al. Transient dynamics of powder spattering in laser powder bed fusion additive manufacturing process revealed by in-situ high-speed high-energy x-ray imaging. Acta Materialia. 2018;**151**:169-180

[43] Wu Z et al. The effect of defect population on the anisotropic fatigue resistance of AlSi10Mg alloy fabricated by laser powder bed fusion. International Journal of Fatigue. 2021;**151**:106317

[44] FLOW-3D® Version 12.0 Users Manual. FLOW-3D [Computer Software]. Santa Fe, NM: Flow Science, Inc.; 2018

[45] Lin R, Wang HP, Lu F, Solomon J, Carlson BE. Numerical study of keyhole dynamics and keyhole-induced porosity formation in remote laser welding of Al alloys. International Journal of Heat and Mass Transfer. 2017;**108**:244-256

[46] General Information on 3D Print Process Simulation/ANSYS Mechanical Workbench Additive 2019R3. Accessed: February 21, 2021

Chapter 4

Framework for Design and Additive Manufacturing of Specialised Multirotor UAV Parts

Petar Piljek, Nino Krznar, Matija Krznar and Denis Kotarski

Abstract

Rapid prototyping technologies have enabled a major step forward in the development of a very wide range of products, especially in the field of mechatronic systems. These technologies are largely related to additive manufacturing (AM), so-called 3D printing which is, in addition to product development, also suitable for the fabrication of mechatronic systems that are not intended for series production. In this chapter, a framework for the AM of specialised multirotor unmanned aerial vehicles (UAVs) parts is proposed and described for three AM technologies—fused deposition modelling (FDM), selective laser sintering (SLS), and stereolithography (SLA). A different approach to parts design is shown where the main problems are addressed and guidelines for parts manufacturing are given. Special emphasis is related to the mechanical characteristics and low weight of the manufactured parts that are merged with carbon fibre segments. The manufactured (printed) parts are mounted in functional assemblies and preliminarily tested.

Keywords: specialised multirotor UAV, additive manufacturing, fused deposition modelling, selective laser sintering, stereolithography

1. Introduction

In the last 10 years, the market for unmanned aerial vehicles (UAVs) in the civil sector has been growing enormously. This was certainly preceded by a period of intensive research that continues to this day, so, an even greater step forward is expected in the future. Technological advances in the design and manufacture of mechatronic system components have enabled many applications from the aspect of automation. The development of control, propulsion, power supply components, and other subsystems has contributed to greater speed of data processing and greater autonomy, which enables the performance of complex flight missions. The development of propulsion components and numerous studies of propulsion configurations have facilitated applications in various sectors, such as precision agriculture [1, 2], surveillance [3], and aerial photography [4]. The application possibilities of UAVs are plentiful in many other sectors, such as transport [5], construction [6], fire protection [7], and more.

The propulsion configuration defines how the aircraft will move in three-dimensional space and it depends on the type of application or mission that the UAV needs to perform. Numerous types of aircraft with various propulsion configurations are used to perform different tasks, activities, and for research and development. In addition to conventional types of UAVs with fixed wings [8, 9] and rotary wings [10–12], a number of hybrid configurations [13, 14] and bioinspired propulsion configurations [15, 16] are being investigated. Fixed-wing aircraft can achieve high speeds and compared to other types, consume less energy to achieve movement, but on the other hand, unable to perform the stationary flight. Generally, they need a runway or special launchpad to be able to take off. Aircraft with rotary wings do not have this problem because they have the ability to take off and land vertically (VTOL), and thus stationary flight and flight at moderate speed. This makes them suitable for missions that require complex manoeuvres and a higher degree of system autonomy. Within the rotary-wing UAV type, there are numerous subtypes of aircraft. It is important to highlight two typical representatives, aircraft with variable pitch propellers, such as helicopter aircraft [17] and multirotor aircraft (multicopter) [18], consisting of N rotors on which fixed-pitch propellers are mounted.

Multirotor type of UAV has greater agility and manoeuvrability, which allows them to perform missions that involve precise and complex movements. On the other hand, they are characterised by high-energy consumption, so it is extremely important to choose the right components and parameters of the system. The most commonly used configuration utilises four rotors (so-called quadrotor) and to a lesser extent the configuration with six (hexarotor), and eight rotors (octorotor). Generally, conventional configurations are characterised by a planar geometric arrangement of an even number of rotors. In addition to conventional purposes, a variety of propulsion configurations makes the multirotor type of UAV suitable for usage as aerial robotic systems. Since this type of application is expected for specialised tasks, there is a need to design custom aircraft and make small series or customised systems. It is also important to save time in the design and production phase and lower production costs compared to conventional manufacturing technologies. Rapid prototyping technologies, such as additive manufacturing (AM), allow the fabrication of assembly parts of such systems [19–21]. Numerous studies have shown the possibilities of rapid prototyping technologies and their application [22, 23].

In this chapter, the framework for design and AM of specialised multirotor UAV parts is presented. In the system design phase, it is necessary to select components and design multirotor UAV based on the purpose of the aircraft. The division into modules (subsystems) allows a greater degree of modularity that leads to a wider range of applications (by fitting the aircraft with different equipment). In the prototyping and production phase, the procedure for making parts using three different AM technologies is described. Depending on the mechanical and other requirements, which are defined in the system design phase, FDM, SLS, and SLA technologies are used within this framework. Professional and hobby 3D printers and related software packages were used in the production process. The procedure was validated for two considered case studies, for a small fully-actuated modular aircraft, and a heavy-lift multirotor UAV. The last part of this chapter presents experimental testing in certain phases of the specialised UAV development, which is necessary for this type of aircraft to be safely used.

2. Multirotor UAV system description

Multirotor type of UAV is classified as rotary-wing UAV, aircraft that are heavier than air and are powered by motors. The ability to take off and land vertically, hover, and fly at moderate speeds, amongst other flight manoeuvres, allows multirotor UAVs to perform complex movements, making them suitable for a wide range of tasks. From a mechanical point of view, the multirotor type of UAV system is described as a rigid body consisting of *N* rotors (propulsion units) that exist in 3D space; hence, it has six degrees of freedom (DOF). Such a multivariable system is mathematically described by a dynamic model with six second-order differential equations. The geometric arrangement of the propulsion subsystem defines the aircraft configuration. To perform missions such as aerial filming, conventional configurations characterised by a planar arrangement of the even number of rotors are generally used. Commercial aircraft for these and similar purposes are mainly quadrotor (quadcopter), hexarotor (hexacopter), and octorotor (octocopter) aircraft. The listed configurations can be in + and × arrangement (layout), such as configurations shown in **Figure 1**.

The design of the aircraft system primarily depends on the purpose, respectively, the mission profile that the aircraft should typically perform. To allow easier analysis of aircraft parameters and design, the aircraft system can be divided into four key subsystems (**Figure 2**). The equipment and payload to be carried by aircraft dictate the choice of parameters and components of other subsystems. The rotors of the propulsion subsystem are mainly electric propulsion units (EPUs) whose central part is a brushless DC (BLDC) motor with a corresponding electronic speed controller (ESC), and a fixed-pitch propeller mounted on a motor rotor. By their rotation, the propellers create aerodynamic forces and moments and directly affect the flight dynamics, which means that the rotors angular velocities are the input variables of the propulsion subsystem. The characteristic of the multirotor UAVs is high-energy consumption, so an energy subsystem must deliver a large amount of energy. In conventional

Figure 1.
Conventional multirotor UAV configurations in ×-layout.

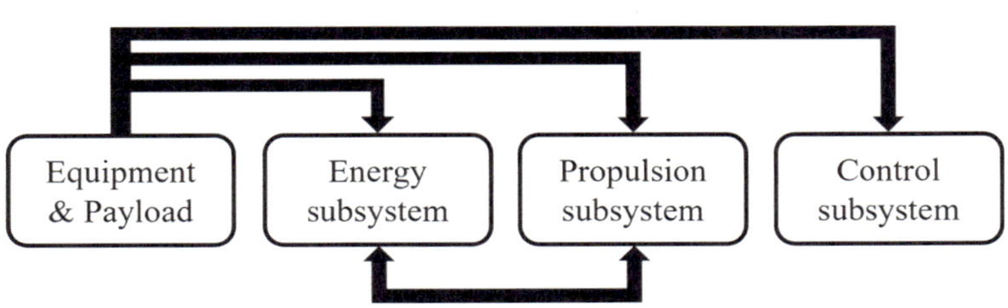

Figure 2.
Multirotor UAV main subsystems.

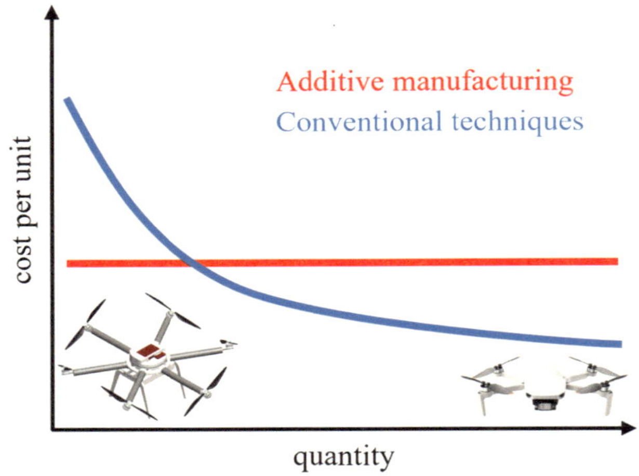

Figure 3.
Cost per unit with respect to quantity for conventional and additive manufacturing technologies.

EPUs, the power subsystem mainly consists of one or more lithium-polymer (LiPo) batteries with associated electronics. The design of the control subsystem or the selection of components primarily depends on the mission or the degree of autonomy that determines the selection of the flight controller, sensors, and other peripheral modules (telemetry, RC, VTx, and others). It follows that the performance of a multirotor type of UAV is determined by the parameters and components of the propulsion and energy subsystems. These two subsystems are interdependent because, for example, as the power of the aircraft increases, the energy demand increases, resulting in a higher mass of the aircraft. The energy requirements of the propulsion subsystem must be taken into account when selecting batteries, which, in turn, depends on the weight and size of the aircraft and the number of EPUs. When designing a system, the ratio of mass and capacity of the battery is one of the key data.

In this chapter, the design of specialised multirotor aircraft is considered, and two case studies are presented through the design, production, and testing phases. Aircraft, such as those used in the case study, cannot be procured in form of commercial aircraft produced in large series. They are produced in small series or even as unique models designed to perform a specialised task. The first case is an experimental modular multirotor (EMMR) UAV with a power of 350–700 W, which has so far been proposed as an engineering educational platform [18]. EMMR can be used as an aerial robotic system since fully-actuated UAV configurations can be assembled. Such a platform represents a suitable engineering educational tool due to the complexity of the system, which requires an interdisciplinary approach in the field of mechanical engineering, electrical engineering, and computing. The second case is a heavy lift aircraft that can be a power of approximately 10–20 kW, depending on the number of rotors. Such an aircraft is considered for use in precision agriculture for smart spraying tasks. In addition to the fact that these aircraft are not commercially available in a form that would allow change of the parameters within open-source software, it is also important to point out that in small series production the cost per unit increases dramatically. For this reason, technologies for rapid prototyping were chosen, mostly AM in which the cost per unit is the same regardless of the number of units produced (**Figure 3**), which is a known fact described in numerous studies [24, 25]. AM is often appropriate for small to medium-sized production series but there is always an inflexion point at which other manufacturing methods become more cost-effective.

3. Additive manufacturing technologies

In this chapter, AM technologies are used for the rapid prototyping and development of specialised multirotor UAVs. In addition to the fact that for small batches AM is cheaper compared to conventional processes, it also significantly shortens the development time by rapid iteration and the possibility of early and often testing many different designs or partial designs with critical features, which further reduce the cost of the final product. Conventional production technologies are much more expensive for small batches due to preparation, tool selection, manufacturing of tools, and other costs. AM, on the other hand, allows the production of parts directly from solid CAD models using software packages, so-called slicers. AM is also suitable for the production of spare parts for damaged aircraft.

There are a large number of low-cost 3D printers on the market, so for low-power multirotor aircraft, parts can be produced very cheaply and quickly. 3D printers may vary greatly in price, size, material, and AM technology used. The paper further considers three AM technologies: FDM, SLS, and SLA. 3D printing uses a wide range of materials, the choice of which is related to AM technology and the purpose of the part. In the case of aircraft parts, plastic materials in the raw form of filament, powder, or resin are mainly used. To determine whether certain materials and AM technologies are suitable for the production of a particular part, the desired strength, stiffness, and weight of the part must be taken into account, but the influence of environmental conditions and the expected duration of the part must also be considered. In addition to the choice of material, the mechanical properties of the part can be alternated and adjusted by changing the printing parameters and the orientation of the printed part. Because parts are fabricated gradually, layer by layer, the inevitable result is the anisotropic properties of printed parts. Better mechanical properties are achieved along with the printing layer and worse in a direction normal to the printing layer. There are many ways in which the mechanical properties of materials can be tested [26–28]. Also, greater precision and greater detailed geometry can be achieved in planes parallel to the print layer where print accuracy is higher. **Table 1** shows the main characteristic of the used 3D printers in combination with the associated software.

AM technology	3D printer	Raw material form	Build volume	Software
FDM	Prusa i3 MK3S+	Continuous thermoplastic filaments	250 × 210 × 210 mm	PrusaSlicer
FDM	Markforged Onyx Pro	Composite base filaments	320 × 132 × 154 mm	Eiger
SLS	Sinterit Lisa Pro	Powder	150 × 200 × 260 mm	Sinterit Studio
SLA	Formlabs Form 3	Resin	145 × 145 × 185 mm	PreForm

Table 1.
Used 3D printers with associated software.

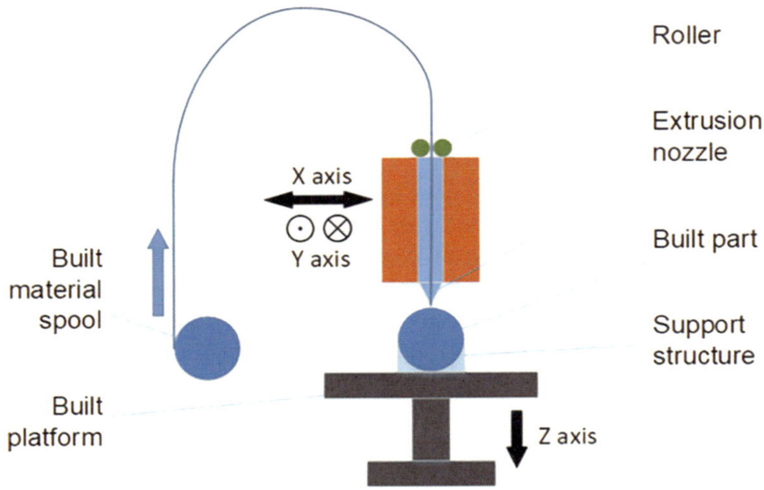

Figure 4.
The principle of operation of FDM technology.

3.1 Fused deposition modelling

Fused deposition modelling (FDM) or known as fused filament fabrication (FFF) is a manufacturing technology in which objects are created by extruding polymer filament onto a built platform through a heated nozzle. There are numerous versions of FDM printers with various price ranges. In this research, Prusa i3 MK3 is used as a low-cost FDM printer where the platform moves in the Y-axis and the nozzle in the X- and Z-axes. When one layer is done, the nozzle will move up vertically to allow a new layer to be applied to the previous one. The thickness of the layer (slice) depends on the print parameters, and in the case of the used Prusa printer, the slices are between 0.05 and 0.30 mm thick [29]. Prior to the AM process, the constructed CAD model must be exported in a compatible file format, such as STL. Such a model is then cut into horizontal slices in a software package (so-called slicer). The paths of the platform and the nozzle are calculated by the software according to the parameters set by the user. In addition to the mentioned layer thickness, which significantly affects the accuracy, some of the other variable parameters are the number of layers in the outer wall and the number of layers at the bottom and top of the part, the percentage and structure of the filling, extrusion speed, and others. Because the next printing layer prints on top of the last one, supporting structures are required to print large overhangs or holes. They are printed together with the part and removed after printing is done. In general, overhangs should be avoided by proper orientation of the part or by using angled overhangs where possible. The most common materials used in FDM technology are ABS, PLA, PC, ASA, PPSF/PPSU, ULTEM, PH-HD. PE-LD, PET, TPU, and others. **Figure 4** shows a working principle of the FDM technology.

3.1.1 Continuous fibre fabrication

In addition to classic FDM technology, devices that can produce parts from composite materials using FDM processes are known as continuous fibre fabrication (CFF). In this paper, Markforged Onyx Pro is used, in which the platform moves in the Z-axis and the nozzle in the X- and Y-axes. Compared to the Prusa printer, it is a much more expensive device but allows 3D printing of composite materials made of plastic matrix and inlaid

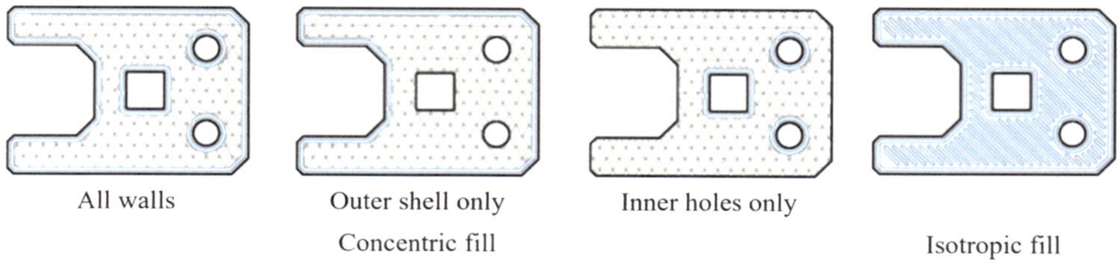

Figure 5.
Fibre reinforcement layout—CFF technology [30].

fibreglass fibres for better mechanical properties and increased lifetime, compared to plastic alone. The strength and stiffness of a fibre-reinforced part can be comparable to aluminium. The software package allows adjustment of the classic print parameters and further adjustment of the composite reinforcements parameters as shown in **Figure 5**.

3.2 Selective laser sintering

The next AM technology considered in the chapter is selective laser sintering, which with the advent of cheaper 3D printer systems allows the application not only for industrial purposes but also for research. The material used in this technology is available in the form of powder that is laser-sintered to create a designed geometry. The powder delivery mechanism consists of two chambers, in the first, there is construction powder that is delivered to the second chamber through rollers and a piston in form of a powder layer. In the second chamber, a layer is precisely sintered to the desired shape utilising laser beams. This technology does not require a support structure, as the unsintered powder provides support to the object under construction. This allows the production of parts of more complex geometry from different types of materials, and it is possible to produce prefabricated assemblies with movable joints. After the production process, further

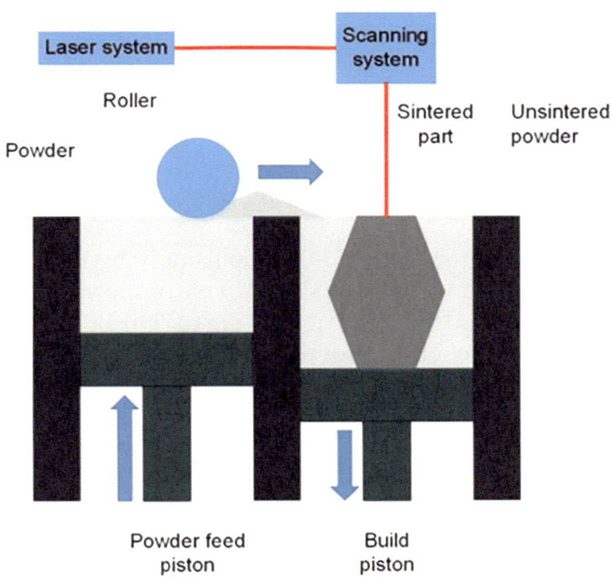

Figure 6.
The principle of operation of SLS technology.

processing of the part or assembly is required to achieve certain mechanical properties of the finishing quality. In this chapter, the SLS system is discussed, which consists of the SLS 3D printer Sinterit Lisa Pro and the associated equipment for the preparation of powder materials (nylon 11, nylon 12, TPU, TPE, and polypropylene) and processing of parts and assemblies. **Figure 6** shows the working principle of the SLS.

3.3 Stereolithography

Stereolithography (SLA) is the first commercially available AM technology developed in 1986 by 3D Systems. With this technology, CAD models are created by curing polymer resin using a laser beam system. With SLA technology, the laser is focussed on a mirror scanning system that cures polymer resin with very high precision. When one layer is cured by laser, the built platform moves upwards in the z-direction and the new layer can be treated. The materials for SLA are thermoset photosensitive resin-shaped polymers. SLA technology makes it possible to achieve high accuracy and a smooth surface, making it the most cost-effective AM technology. Compared to the previously considered technologies, SLA parts have poorer mechanical properties; therefore, SLA technology is not recommended for structurally loaded parts. **Figure 7** shows the scheme of the SLA procedure.

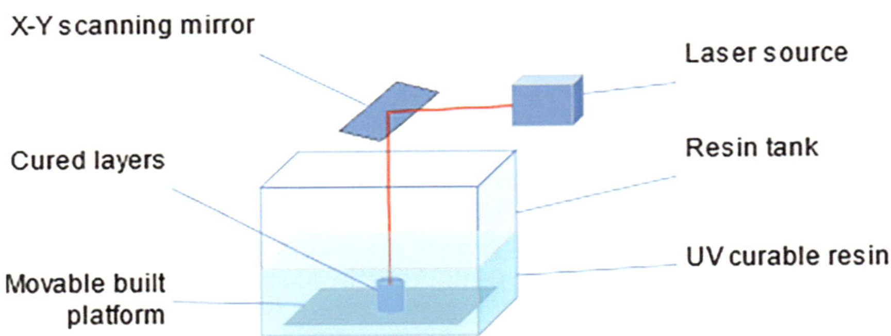

Figure 7.
The principle of operation of SLA technology.

4. Framework for additive manufacturing of specialised UAV parts

The design of the multirotor type of UAV propulsion subsystem is considered and the additive manufacturing framework is shown. This framework can also be used for rapid prototyping of parts from carbon fibre plates. The process of making parts is presented for two experimental aircraft that can be used for specialised purposes, such as performing tasks involving complex and precise movements and in tasks involving the transfer of heavy cargo.

4.1 Propulsion subsystem design considerations

The propulsion subsystem is defined by the parameters of the geometric arrangement and characteristics of the EPUs. A suitable fixed-pitch propeller is mounted on the rotor of the outrunner BLDC motor (**Figure 8**). The basic parameter of a propeller is its diameter. As the diameter of the propeller increases, the angular velocity of the

Figure 8.
Electric propulsion unit of multirotor type of UAV [31].

motor rotor decreases. The motor is defined by a motor velocity constant kV. Motors with a lower motor constant are used in combination with larger diameter propellers and are driven at higher voltages. The ESC is responsible for starting the motor and, depending on the control signal, controls the motor speed. The EPUs are connected to one or more LiPo batteries of the appropriate number of cells and capacity.

The motor stator must be connected to the aircraft assembly which consists of a central part and the rotor arms. Propulsion assembly design is the most complex part of the overall design in terms of the mechanical properties that assembly parts should possess. The aircraft can be used in a wide range of powers, from a few tens of watts to several tens of kilowatts. It is necessary to choose materials and technologies concerning the selected propulsion components. **Figure 9a** shows the stator geometry which is important from the aspect of mounting the motor to the aircraft assembly. **Figure 9b** shows the characteristics of the propulsion unit considered in the case of a heavy-lift aircraft.

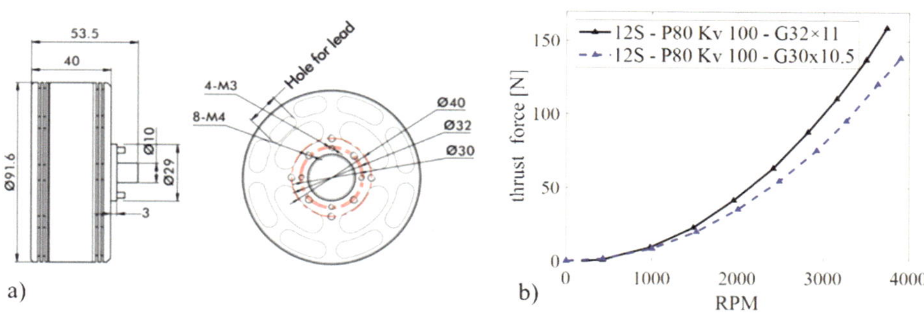

Figure 9.
Electric propulsion unit: (a) BLDC motor geometry [31]; (b) characteristics.

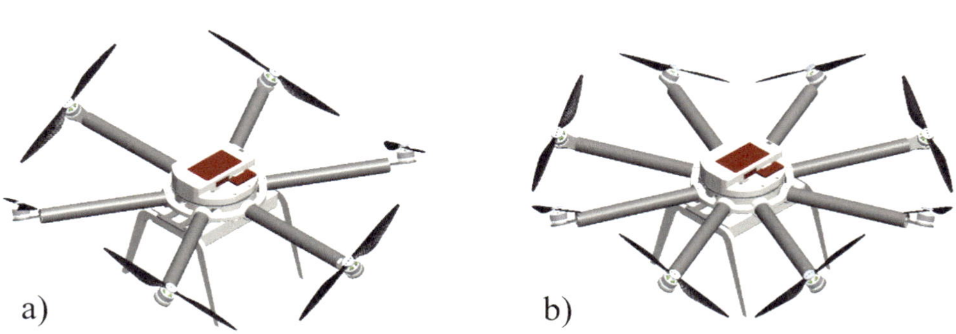

Figure 10.
Fully-actuated multirotor configurations with passively tilted rotors: (a) PTX6; (b) PTX8.

The configuration of the multirotor UAV is defined by the geometric arrangement of the rotors. Mostly conventional configurations with a planar rotor layout are commercially available. It is possible to select configuration parameters that will result in an increased degree of actuation, which potentially allows the performance of complex tasks in the field of aerial robotics. A fully-actuated aircraft with passively tilted rotor arms are considered in this research (**Figure 10**).

4.2 Additive manufacturing procedure

A framework for the production of parts for specialised multirotor UAVs using additive manufacturing is presented. It consists of an aircraft design stage in which various software packages can be used for the needs of 3D modelling of parts and

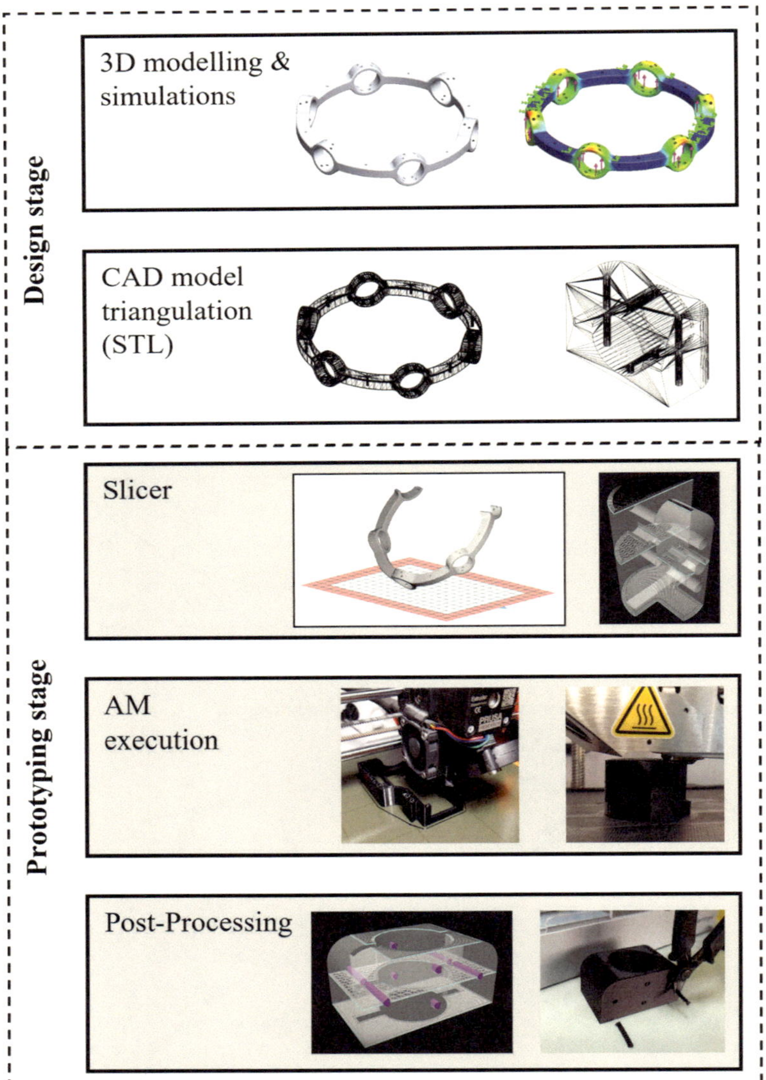

Figure 11.
Additive manufacturing procedure.

assemblies, and also for simulations. In this research, the SOLIDWORKS software package is used in the design stage. After the process of creating a model is done, triangulation of the 3D CAD model is performed and the model is exported into an STL format. In the prototyping stage, it is necessary to adjust the parameters of the 3D print in accordance with the selected AM technology using associated software, the so-called slicer. The next step is the execution of the g-code by which the given parts are produced. After finishing the print, the parts need to be post-processed (**Figure 11**).

4.3 Experimental verification

Manufactured parts of specialised multirotor UAVs are connected together with other components into functional assemblies. Through the prototyping phase, different test phases were conducted for the two aircraft based on propulsion units with the parameters given in **Table 2**. By assembling and testing individual subsystems, potential design errors can be identified, and improvements offered.

The control subsystem of the experimental aircraft is based on the open-source Pixhawk FC. To operate a fully-actuated aircraft, custom firmware has been developed.

Multirotor configuration	BLDC motor	Propeller	ESC
PTX6 D = 500 mm	MN1806 1400 Kv	CF7024 d = 7″	Air 10A 3S
X4 D = 1500 mm	P80 100 Kv	G32x11 d = 32″	Flame 80A 12S

Table 2.
Considered multirotor configuration main parameters.

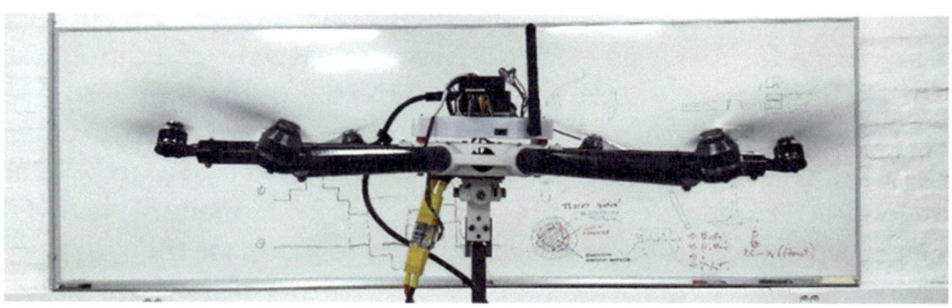

Figure 12.
Experimental testing of PTX6 configuration in case of attitude control.

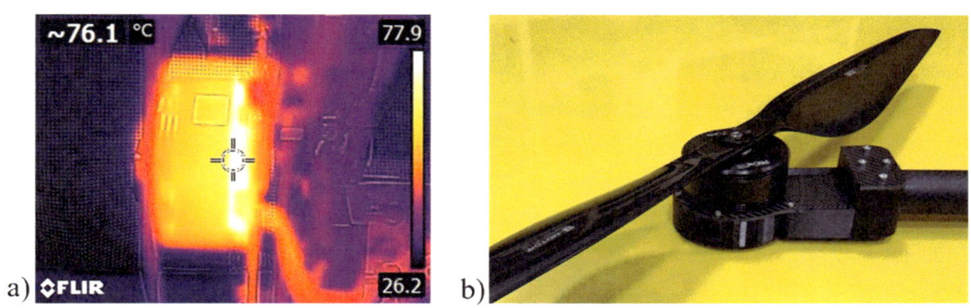

Figure 13.
Heavy-lift aircraft propulsion: (a) EPU testing; (b) EPU assembly.

Figure 12 shows the indoor testing phase where attitude control experiments were conducted. Indoor testing provides a safe way to set the basic parameters of the control subsystem and set up and test all safety elements. It is also possible to tune the parameters of the control algorithm. After the indoor phase, the remote control of the aircraft was tested in two cases that differ by control inputs from the RC transmitter. The first case is represented with conventional control inputs (thrust, roll, pitch, and yaw), while in the second, control inputs were three forces and yaw moment with respect to body axes.

For the second experimental aircraft, the propulsion unit was tested in different operating regimes at the full power range. Characteristics were obtained (**Figure 9b**) and other parameters, such as heating, were monitored (**Figure 13a**). Given the power of the aircraft, the described framework is used in a wider range of rapid prototyping, which includes cutting carbon plates, which together with printed parts and prefabricated tubes form the rotor arm assembly (**Figure 13b**). In the coming period, it is planned to assemble the propulsion subsystem into a functional assembly so that tests can be carried out as in the case of the first experimental aircraft.

5. Conclusion

This chapter demonstrates the application of three different AM technologies for the development of customised parts for the specialised multirotor UAVs—fused deposition modelling (FDM), selective laser sintering (SLS), and stereolithography (SLA). Special purpose multirotor UAVs are often produced in small series, with the option of personalization and modular design. In the case of prototyping or individual production, conventional manufacturing technologies are too expensive and not flexible enough to be able to make parts quickly and put them into exploitation. AM offers new possibilities for rapid development of UAV multirotor reducing costs and time of research, development, and production. To take full advantage of AM, a new design approach for AM is needed to achieve lightweight and durable structures of UAV parts. Preliminary tests have shown that the use of the proposed AM technologies is very promising in terms of designing parts of specialised aircraft, as many factors (i.e., geometry, strength, firmness, and weight) often have to be changed and adjusted during the design process. In future work, it is planned to use AM technologies to make parts of other aircraft subsystems and to integrate them in the overall multirotor UAV system. Furthermore, the oncoming tests of mechanical properties are expected to have a great significance for frame structure optimisation.

Acknowledgements

This research was funded by European Regional Development Fund, Operational programme competitiveness and cohesion 2014–2020, as part of the call for proposals entitled “Investing in science and innovation—first call,” grant number KK.01.1.1.04.0092.

Appendices and nomenclature

AM	additive manufacturing
UAV	unmanned aerial vehicle

FDM	fused deposition modelling
SLS	selective laser sintering
SLA	stereolithography
VTOL	vertically take-off and land
DOF	degrees of freedom
EPU	electric propulsion unit
BLDC	brushless direct current
ESC	electronic speed controller
FFF	fused filament fabrication
CFF	continuous fibre fabrication
LiPo	lithium-polymer

Author details

Petar Piljek[1], Nino Krznar[1], Matija Krznar[1] and Denis Kotarski[2]*

1 Faculty of Mechanical Engineering and Naval Architecture, University of Zagreb, Zagreb, Croatia

2 Karlovac University of Applied Sciences, Karlovac, Croatia

*Address all correspondence to: denis.kotarski@vuka.hr

References

[1] Yinka-Banjo C, Ajayi O. Sky-farmers: Applications of unmanned aerial vehicles (UAV) in agriculture. In: Dekoulis G, editor. Autonomous Vehicles. London: IntechOpen; 2020. DOI: 10.5772/intechopen.89488

[2] Zhang C, Valente J, Kooistra L, Guo L, Wang W. Orchard management with small unmanned aerial vehicles: A survey of sensing and analysis approaches. Precision Agriculture. 2021;**22**:2007-2052. DOI: 10.1007/s11119-021-09813-y

[3] Templin T, Popielarczyk D, Kosecki R. Application of low-cost fixed-wing UAV for inland lakes shoreline investigation. Pure and Applied Geophysics. 2018;**175**:3263-3283. DOI: 10.1007/s00024-017-1707-7

[4] Stanković M, Mirza MM, Karabiyik U. UAV forensics: DJI mini 2 case study. Drones. 2021;**5**. DOI: 10.3390/drones5020049

[5] Kotarski D, Piljek P, Kasać J. Design considerations for autonomous cargo transportation multirotor UAVs. In: Găiceanu M, editor. Self-Driving Vehicles and Enabling Technologies. London: IntechOpen; 2021. DOI: 10.5772/intechopen.95060

[6] Palacios AT, Cordero JM, Bello MR, Palacios ET, González JL. New applications of 3D SLAM on risk management using unmanned aerial vehicles in the construction industry. In: Dekoulis G, editor. Drones-Applications. London: IntechOpen; 2018. pp. 97-118. DOI: 10.5772/intechopen.73325

[7] Yuan C, Liu Z, Zhang Y. Aerial images-based forest fire detection for firefighting using optical remote sensing techniques and unmanned aerial vehicles. Journal of Intelligent & Robotic Systems. 2017;**88**:635-654. DOI: 10.1007/s10846-016-0464-7

[8] Escobar-Ruiz AG, Lopez-Botello O, Reyes-Osorio L, Zambrano-Robledo P, Amezquita-Brooks L, Garcia-Salazar O. Conceptual design of an unmanned fixed-wing aerial vehicle based on alternative energy. International Journal of Aerospace Engineering. 2019;**2019**:1. DOI: 10.1155/2019/8104927

[9] Mohammed Z, Khaled F, Mohamed A. Adaptive integral high-order sliding mode for a fixed wing aircraft. In: Bartoszewicz A, editor. Recent Developments in Sliding Mode Control. London: IntechOpen; 2017. DOI: 10.5772/67580

[10] Cai G, Feng L, Chen BM, Lee TH. Systematic design methodology and construction of UAV helicopters. Mechatronics. 2008;**18**:545-558. DOI: 10.1016/j.mechatronics.2008.05.011

[11] Jung S, Jo Y, Kim YJ. Flight time estimation for continuous surveillance missions using a multirotor UAV. Energies. 2019;**12**. DOI: 10.3390/en12050867

[12] Kotarski D, Piljek P, Kasać J, Majetić D. Performance analysis of fully actuated multirotor unmanned aerial vehicle configurations with passively tilted rotors. Applied Sciences. 2021;**11**. DOI: 10.3390/app11188786

[13] Saeed AS, Bani Younes A, Cai C, Cai G. A survey of hybrid unmanned aerial vehicles. Progress in Aerospace Sciences. 2018;**98**:91-105. DOI: 10.1016/j.paerosci.2018.03.007

[14] Zong J, Zhu B, Hou Z, Yang X, Zhai J. Evaluation and comparison of hybrid wing VTOL UAV with four different electric propulsion systems. Aerospace. 2021;**8**. DOI: 10.3390/aerospace8090256

[15] Hui Z, Zhang Y, Chen G. Aerodynamic performance investigation on a morphing unmanned aerial vehicle with bio-inspired discrete wing structures. Aerospace Science and Technology. 2019;**95**:1. DOI: 10.1016/j.ast.2019.105419

[16] Ryu SW, Lee JG, Kim HJ. Design, fabrication, and analysis of flapping and folding wing mechanism for a robotic bird. Journal of Bionic Engineering. 2020;**17**:229-240. DOI: 10.1007/s42235-020-0018-3

[17] Alvarenga J, Vitzilaios NI, Valavanis KP, Rutherford MJ. Survey of unmanned helicopter model-based navigation and control techniques. Journal of Intelligent & Robotic Systems. 2015;**80**:87-138. DOI: 10.1007/s10846-014-0143-5

[18] Kotarski D, Piljek P, Pranjić M, Grlj CG, Kasać J. A modular multirotor unmanned aerial vehicle design approach for development of an engineering education platform. Sensors. 2021;**21**. DOI: 10.3390/s21082737

[19] Esakki B, Mathiyazhagan S, Moses M, Rao KJ, Ganesan S. Development of 3D-printed floating quadrotor for collection of algae in remote water bodies. Computers and Electronics in Agriculture. 2019;**164**:1. DOI: 10.1016/j.compag.2019.104891

[20] Junaid AB, Diaz De Cerio Sanchez A, Bosch JB, Vitzilaios N, Zweiri Y. Design and implementation of a dual-axis tilting quadcopter. Robotics. 2018;7. DOI: 10.3390/robotics7040065

[21] Perez-Jimenez M, Montes-Grova MA, Ramon-Soria P, Arrue BC, Ollero A. POSITRON: Lightweight active positioning compliant joints robotic arm in power lines inspection. In: Proceedings of the IEEE International Conference on Unmanned Aircraft Systems (ICUAS '20); 1-4 September 2020; Athens. Greece: IEEE; 2020. pp. 729-736

[22] Bird DT, Ravindra NM. Additive manufacturing of sensors for military monitoring applications. Polymers. 2021;**13**. DOI: 10.3390/polym13091455

[23] Garcia Rubiales FJ, Soria PR, Arrue BC, Ollero A. Soft-tentacle gripper for pipe crawling to inspect industrial facilities using UAVs. Sensors. 2021;**21**. DOI: 10.3390/s21124142

[24] Thomas DS, Gilbert SW. Costs and Cost Effectiveness of Additive Manufacturing A Literature Review and Discussion. NIST Special Publication 1176. DOI: 10.6028/NIST.SP.1176

[25] Goh GD, Agarwala S, Goh GL, Dikshit V, Sing SL, Yeong WY. Additive manufacturing in unmanned aerial vehicles (UAVs): Challenges and potential. Aerospace Science and Technology. 2017;**63**:140-151. DOI: 10.1016/j.ast.2016.12.019

[26] Brischetto S, Ferro CG, Maggiore P, Torre R. Compression tests of ABS specimens for UAV components produced via the FDM technique. Technologies. 2017;**5**. DOI: 10.3390/technologies5020020

[27] Păcurar R, Păcurar A, Petrilak A. The influence of build orientation on the mechanical properties of medical implants made from PA 2200 by selective laser sintering. MATEC Web of Conferences. 2017;**112**:1. DOI: 10.1051/matecconf/201711203009

[28] García-Domínguez A, Claver J, Camacho AM, Sebastián MA. Considerations on the applicability of test methods for mechanical characterization of materials manufactured by FDM. Materials. 2020;**13**. DOI: 10.3390/ma13010028

[29] Prusa i3 mk3s. Available from: https://www.prusa3d.com/product/original-prusa-i3-mk3s-3d-printer-3/#Specs [Accessed: 04 January 2022]

[30] Design Guide for 3D Printing with Composites. Avaivable from: https://static.markforged.com/downloads/CompositesDesignGuide.pdf [Accessed: 05 January 2022]

[31] Electric Propulsion Unit P80. Available from: https://store.tmotor.com/goods.php?id=784 [Accessed: 15 December 2021]

Chapter 5

Internal Microchannel Manufacturing Using Stereolithographic 3D Printing

Bastián Carnero, Carmen Bao-Varela, Ana Isabel Gómez-Varela and María Teresa Flores-Arias

Abstract

Internal channels are one of the most interesting structures to implement in microfluidics devices. Unfortunately, the optical technologies typically used in microfluidics, such as photolithography or reactive ion etching, are unable to generate these structures by only allowing surface structuring. Stereolithographic 3D printing has emerged as a very promising technology in internal microchannel manufacturing, by allowing a layer-by-layer structuring in volume performed by a laser that photopolymerises a liquid resin. Recent advances in laser technologies have reached resolutions of tens of micrometres. The high resolution of this type of printer, which a priori would allow the fabrication of channels of the same dimensions, may pose a problem by impeding the evacuation of uncured resin. In this chapter, the compromise between size and resin evacuation will be evaluated to find the optimal diameter range in which unobstructed and accurate microchannels can be obtained.

Keywords: stereolithography, 3D printing, microchannel, microfluidics, laser

1. Introduction

The 3-dimensional (3D) printing has recently become one of the most promising and ground-breaking manufacturing techniques [1–3], allowing to produce highly detailed structures, following simple and systematic steps without the need of the very expensive equipment of traditional technologies that normally require the use of cleaning rooms in large facilities. The 3D printing has facilitated the access to complex processes of manufacturing to a lot of researchers and many and varied industries [4]. Among others, the microfluidics field is a clear beneficiary from the role that 3D printing plays in the microfabrication processes [5], where techniques such as reactive ion etching (RIE) [6] and photolithography [7, 8] that produce a significant polluting chemical waste are predominant.

In addition to its multiple applications in chemistry, engineering or sensing, microfluidics is of great interest in medicine and pharmacology, where one of the challenges is to manufacture complex devices capable of mimicking physiological structures [9–11], such as vessels, veins and arteries, where novel drugs can be tested

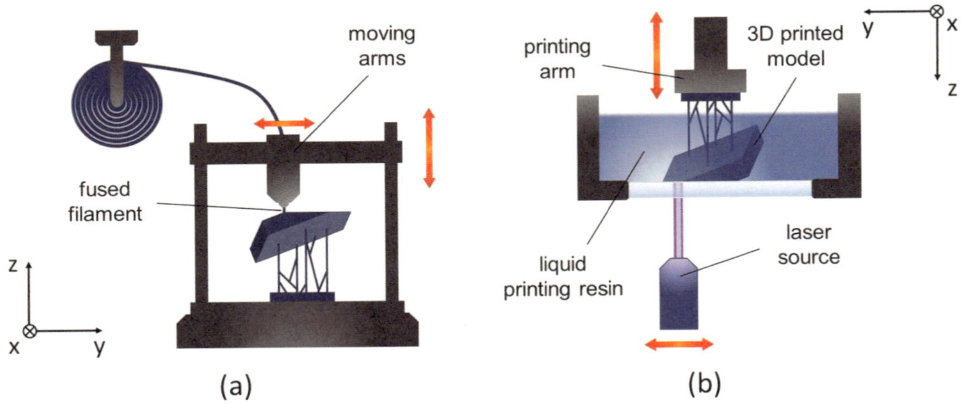

Figure 1.
Popular 3D printing technologies: (a) fused deposition modelling (FDM) and (b) stereolithography (SLA).

under static and flow conditions (dynamic regime). These studies, much closer to reality than the studies carried out by traditional methods, involving testing in wells (static regime), could decrease the animal experimentation needed for testing drugs before the patient dispensation. All these devices are made up of different kinds of microchannels, capable of guiding small amounts of liquid samples. To be able to fabricate these devices, new technologies are required to manufacture them in a repeatable and accurate way. The 3D technology emerges as a promising one, since, it allows to achieve in an easy and fast way, microchannels with very high resolutions with simple procedures; to select different geometries for the microchannel profile (circular, rectangular, triangular...) and to create channels on complex surfaces in 3D or even internally.

Currently, two 3D printing technologies outstand above the rest [12, 13]: fused deposition modelling (FDM) [14, 15] and stereolithography (SLA) [16, 17]. FDM printers are based on the extrusion of a heated polymeric filament fused, that forms consecutive layers of a piece (**Figure 1a**). SLA printers use photopolymerisation to selectively cure a liquid resin contained in a tank (**Figure 1b**), manufacturing the model in a precise layer by layer process.

Both technologies are widely used given their versatility and efficiency, but SLA offers the highest accuracies [18]. Given the high quality of the surfaces fabricated by SLA printers, a variety of biocompatible materials suitable for its use with this equipment have emerged, increasing the potential biological applications to be used for [19–22]. There are many examples that show the perspective of SLA printers for complex microfluidic devices fabrication regarding biological applications, thus, making them an option to be used by researchers focused on 3D printing of reliable accurate and biologically solvent microfluidic devices. However, some technical aspects must be considered to optimise the printing results.

1.1 Theoretical fundaments of stereolithography

The polymerisation of photosensitive resins is mainly governed by two parameters [23]: penetration depth of the curing light and the minimum energy required for polymerisation. The penetration of light follows the well-known Beer-Lambert law of exponential light absorption given by:

$$P_z = P_0 e^{-z/D_P} \tag{1}$$

being P_Z the light power measured at a depth z from the surface; P_0, the power at the surface; and D_P, the depth reached when light intensity decreases by a factor $1/e$ of the surface intensity. Note that D_P is a factor that depends on the resin composition, which determines its absorbance characteristics (dispersion and absorption) [24]. Power terms can be rewritten in terms of energy (so z will become the cure depth when the appropriate amount of light is provided) to obtain the working curve equation for SLA 3D printers:

$$C_D = D_P \ln\left[\frac{E_0}{E_c}\right] \tag{2}$$

where C_D is the depth/thickness at which the light energy is sufficient to convert the liquid resin into a gel; E_0 is the energy of light at the surface; and E_C is the critical energy necessary to initiate photopolymerisation. According to the Beer-Lambert law, the exposed light intensity reaches its maximum value (E_{MAX}) at the surface of the resin, and decreases exponentially as light penetrates through the resin due to the attenuation of the absorbing medium.

In the resin, the photopolymerised volume increases with the ultraviolet (UV) irradiation until the resin reaches to the gel point, where it transforms from liquid to solid-state. D_P and E_C are parameters that depend on the chemical characteristics of the resins and can be determined by drawing a semi-log plot of C_D vs. E_0 obtaining a straight-line curve with slope D_P and an x-intercept of E_C [25]. Once D_P and E_C are known, it is possible to optimise printing process choosing properly the exposure parameters and achieving the designed piece properties. This is the key for obtaining good results with a high-resolution SLA printing, where minimising the thickness of the deposited and light cured layer to achieve the maximum detail is critical.

In most SLA printers, the light source used to perform photopolymerisation is a laser, so the XY resolution is given by the size of the laser spot on the surface. Knowing the aforementioned parameters, the user or printer manufacturer can choose the proper parameters of light exposure (scan speed, power) to optimise the curing conditions and achieve the best resolution for the final device. Another determining factor is the minimum Z-step allowed by the printing arm, which gradually raises the piece from the bottom of the tank, that determines the corresponding layer thickness for each resin (see **Table 1**).

Printing resin	Clear	Model	Tough	Amber	Flexible	Elastic	Dental
Z-step (μm)	25	25	50	50	50	100	100
Washing time (min.)	15 + 5	10	10 + 10	20	10 + 10	10 + 10	20
Biocompatibility	✓	×	×	✓	×	×	✓
Curing temperatures (°C)	*60*	*60*	*70*	*70*	*60*	*60*	*80*
Curing time (min.)	30	60	60	30	60	20	20
Transparency	✓	×	×	✓	✓	✓	✓

Table 1.
Manufacturer characteristics of the resins used in this chapter.

Finally, one of the most important aspects to be analysed for obtaining suitable internal channels is the orientation of the designed device, thus, a deep study of the influence of the inclination of the device to be fabricated in the process of photopolymerisation is necessary in order to determine the configurations that provide better results. We have to realise that the printer will slice the piece in a series of layers parallel to the base so that if the original configuration is rotated, these layers will change together with the areas that will be cured. Hence, objects with high surface detail should be printed with an orientation that helps the accurate curing of the layers. It also happens in the case of internal channels, where a proper angle could favour the full evacuation of the wastes of liquid resin from its interior, avoiding clogging.

In this work, a study of the performance of an SLA 3D printer in microfluidic devices is presented. For this purpose, an annular piece with a series of internal channels of different diameters and angles will be designed and manufactured. The dependence on the printing orientation of the device in the results will be evaluated. The study will be made for seven different commercial printing resins.

2. Materials and methods

2.1 3D printing

A Form 3B printer (Formlabs, Somerville, Massachusetts) is used to print the devices to be studied. This printer features a new technology called Low Force Stereolithography, a step further in SLA printers designed to reduce the manufacturing stress that pieces undergo during printing. In brief, this technology combines a galvanometric system with a series of mirrors to grant an incidence of the laser beam (λ = 405 nm, P = 250 mW) perpendicular to the resin tank, whose base will be made of a flexible material capable of deforming when the piece is pushed on it. In this way, the accuracy of the 3D printed structures is improved, as a much more uniform deposition of the laser energy is ensured.

It is well known that the printing orientation will determine the features of the printed devices. Typically, suppliers recommend using 45° as printing orientation in order to optimise the process. Although this recommendation is useful for superficial structures, we realise that for internal channels, the evacuation of uncured resin can produce obstructed lumens [19]. To test the influence of the printing angle on the ability to create internal channels with good quality, a quarter annulus piece was designed (**Figure 2a**) containing seven internal channels oriented at 0°, 15°, 30°, 45°, 60°, 75° and 90° and printed (**Figure 2b**). This study was performed four times for each resin selected, varying the diameter of the internal channels each time. These pieces can be identified in **Figure 2a** as A, B, C and D, corresponding to microchannels with diameter of 250, 500, 1000 and 1500 μm, respectively.

2.2 Data acquisition

The measurement of the diameter was performed using a Nikon MM 400 metallurgic microscope (Nikon Instruments Europe B.V., Amsterdam, The Netherlands), that performs measurements in real time (**Figure 3a**), and an analysis NIS-Elements Nikon software (Nikon Instruments, Melville, USA), by adjusting a measurement circumference (**Figure 3b**) that the software allows to move and modify over the image. The channels were illuminated in transmission light configuration that allowed us to

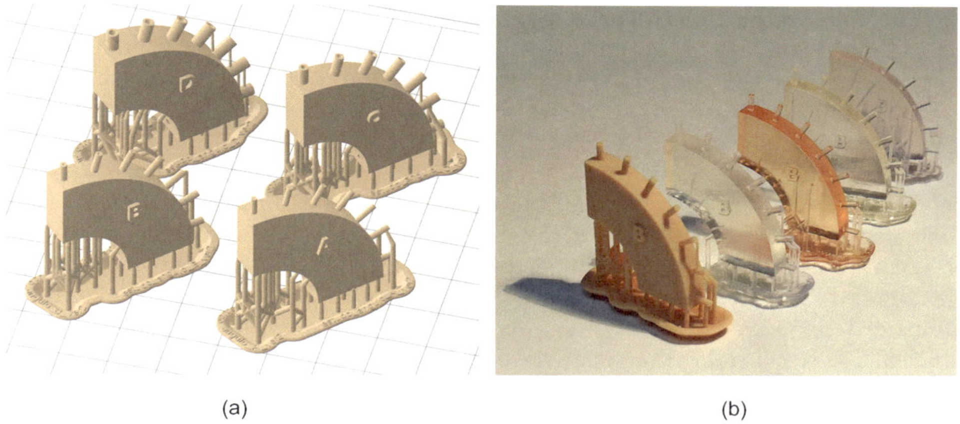

Figure 2.
Picture of some 3D pieces: (a) image of the design used to study the formation of internal channel when varying printing angles and internal diameters, (b) picture of selected annulus printed for different resins, with theoretical internal diameters of 500 μm. From the fore to the ground: Model, Clear, Amber, Dental and Flexible resin. Scaffolding supporting the structures is shown.

Figure 3.
(a) Experimental configuration used to measure the internal channels of the quarter annulus using a microscope. White arrows point channels not fully formed, printed at 0° and 15°. (b) Microscope image of the end of a channel printed using Model resin, at an angle of 75° and a theoretical diameter of 1000 μm. The picture was taken with a 5X microscope objective.

measure the lumen of each one. Images were acquired using a LU Plan Fluor objective (Nikon Instruments, Melville, USA) with 5X magnification and a CCD camera Nikon DS-FI2 (Nikon Instruments, Melville, USA). Five measures were performed for each channel, obtaining a geometric mean and a standard deviation that will be presented in Section 3. Images of longitudinal sections of the microchannel internal surfaces were obtained with a 3D optical profilometer S neox (Sensofar Metrology, Terrassa, Spain) working in confocal mode.

2.3 Materials

Seven printing resins made by Formlabs for the Form 3B printer were studied: Dental LT V1, BioMed Amber V1, Elastic 50A V1, Clear V4, Model V2, Tough 2000 V1 and Flexible 80A V1. As introduced in Section 1.1, one of the critical factors to obtain high accuracy results with SLA printers is the Z-step of the printing arm allowed by every resin. Thus, the minimum Z-step was selected for each of them, as indicated in **Table 1**. Some of the used resins are even biocompatible (**Table 1**), which increases their potential applications.

After printing, it is necessary to post cure the resin pieces in a two-step process, to improve their mechanical aspects and superficial finishing. This process starts with a wash of the part in isopropanol >90% in the Form Wash tank (Formlabs, Somerville, Massachusetts), in one (Model, Amber and Dental) or two cycles (Clear, Tough, Flexible and Elastic), during times indicated in **Table 1**. The pieces are then left to dry and placed in the UV Form Cure chamber (Formlabs, Somerville, Massachusetts), which allows to control the temperature and is also provided with LEDs emitting at 405 nm. Curing temperatures and curing times can be consulted in **Table 1**.

3. Results and discussion

The manufacturing of internal channels with a continuous and unobstructed lumen is one of the main challenges for actual SLA printers, because of their many applications in microfluidics [16, 17]. The fabrication of cavities in a bulk with a proper lumen is a very difficult process, since the photopolymerisation of each layer is sustained by the previous one, so the evacuation of the non-polymerised resin can be tedious.

In many cases, the goal of obtaining unobstructed channels goes against the need for the printer to introduce scaffolds in the largest cavities, causing that internal channel collapse if some supports are not used during the printing. In addition, the own resolution of the printer can act as a limiter for very small channels, which do not have a structural challenge. In order to properly establish the dimensional limit between small channels and large cavities and to study the dependence of the internal channel performance on the diameter and angle of the printer, quarter annulus crossed by internal channels (**Figure 2**) were printed for each resin and the experimental diameters were measured as indicated in Section 2.2.

From the obtained results, three printing regimes can be defined. In the case of channels with small diameters (250 μm), no channel was fabricated for any resin at any angle, so no data can be presented. It can be concluded that, for these sizes, the formation of internal cavities in this range is not possible due to its small size, which prevents the correct evacuation of the resin. This implies that, the resolution for structures inside the printed piece is lower than the resolution for external ones, as structures of this size could be formed if they were made on the surface [19].

Next, for 500–1000 μm in diameter (medium diameters), channels begin to be formed (see **Figure 4a** and **b**) as will be detailed below. The bottom of these channels has been measured using the experimental configuration presented in **Figure 3**. We defined the accuracy as the ratio between the printed and theoretical designed diameter, in percentage. The tendency observed is an increase of experimental diameters as the printing angle increases, for a theoretical fixed value. For channels of 500 μm in diameter (**Figure 4a**), Amber and Dental resins provide the best results, almost reaching a 100% accuracy for an angle of 90°. In addition, for angles greater than 60°, they are all above 80% accuracy, together with Model resin. For lower values of the angles, the channels are narrower than those designed and are more incomplete (longitudinally) as the angle decreases, so for 15°, only Amber

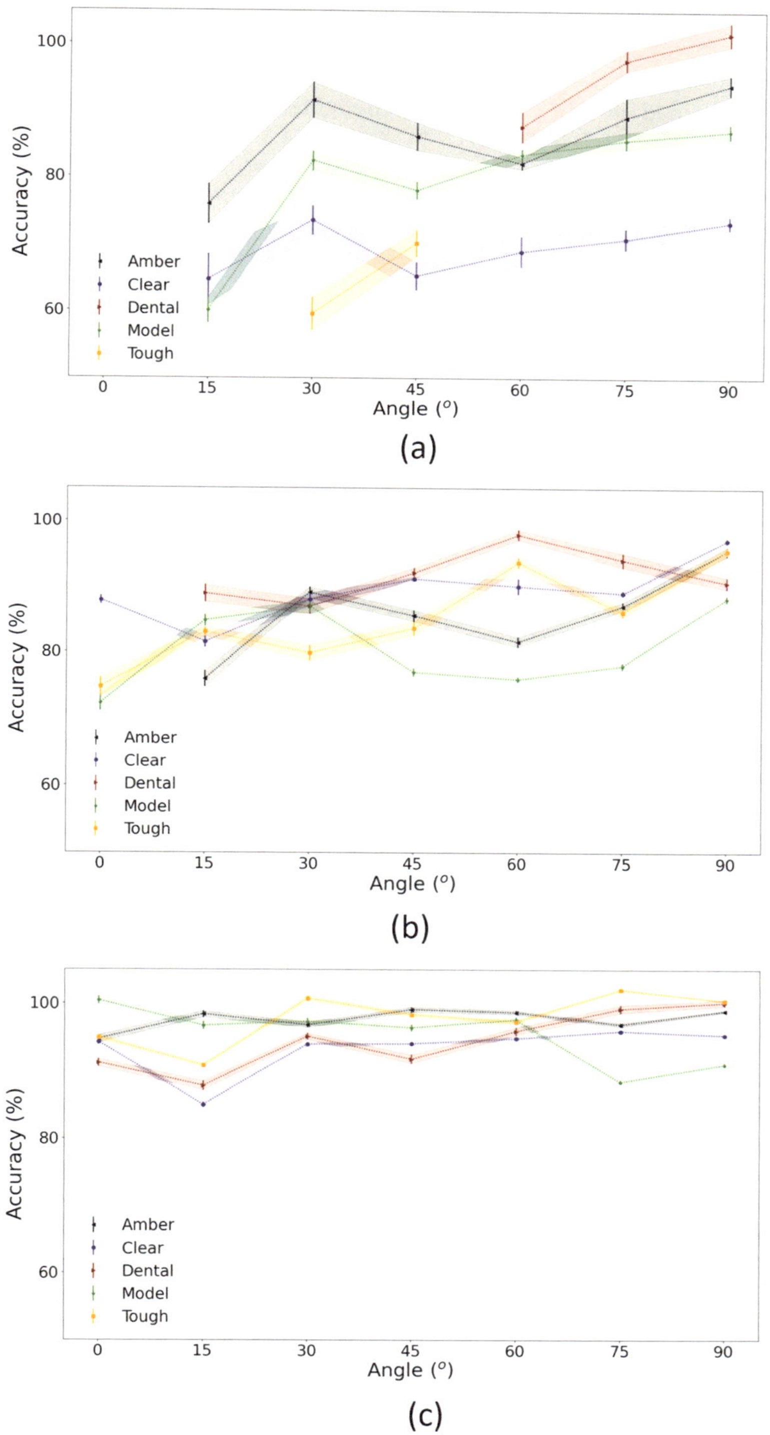

Figure 4.
Accuracy of the printing for the internal channels with diameters of (a) 500 μm, (b) 1000 μm and (c) 1500 μm in diameter, respectively. The error bars represent the standard deviation of the accuracies, and the area between the errors has been filled to facilitate the interpretation of the graphs.

Clear and Model resins form channels and for 0°, none. Longitudinally, Clear and Dental only form complete channels for 90° while Amber resin enables the formation of complete channels for 60°, 75° and 90°. For other values, the channels are not completely formed, although the unobstructed length of the channel increases as the angle increases (see **Figure 3a**).

When channels of 1000 μm in diameter (**Figure 4b**) are fabricated, the printing accuracy suffers a global increase, being always above 70% for every studied angle. As the angle increase, an improvement in the precision is observed, and from 45°, all resins show an accuracy of more than 80% (except for Model, which shows a more irregular trend). The best results are obtained for 90°, where all the resins are above 90%, being the Model resin the exception, reaching an 88%.

In the case of channels with 1500 μm in diameter (wide channels), an 85% on accuracy is achieved for all channels at every studied angle (**Figure 4c**). The length of the channels increases until they form completely (unobstructed) at 45° for Clear resin and at 15° for Amber and Dental resin. For greater angles, complete channels are formed for these resins. For these diameters, results are particularly suitable for angles greater than 60° degrees, where all resins show a printing accuracy greater than 95%, being the exception again the accuracy of Model resin, which is much closer to 90%. Therefore, internal channel with wide diameters allows to fabricate internal cavities for any angle and do not need scaffolding inside. Note that, in the case of the Tough and Model resin, the length of the channels cannot be evaluated by naked eye due to their opacity.

Channels fabricated at 45° and 500 μm in diameter were chosen for inspecting the internal surface of unobstructed channels. In particular, Tough, Clear and Model resins were selected to be analysed because of their different properties (Z-step, biocompatibility, transparency...). **Figure 5** shows confocal images of longitudinal sections of the channels, where it can be observed that semi-circular designed profile is properly translated to the printed pieces.

By comparing the confocal images of the Tough (**Figure 5a**) vs. Clear and Model resins (**Figure 5b** and **c**), a decrease of the surface waviness with the Z-step is observed. The smoothest profile was achieved using the Model resin (**Figure 5c**).

From the previous analysis, we realise that the angle of impression is critical and has a major influence in preventing (**Figure 6a**) or favouring (**Figure 6b**) the formation of internal channels, so a larger angle (closer to 90°) is observed as the most suitable for channels to form properly and to have dimensions closer to those designed. The other key parameter found in this study is the diameter. As we have seen, a larger diameter allows results to be obtained with a higher resolution.

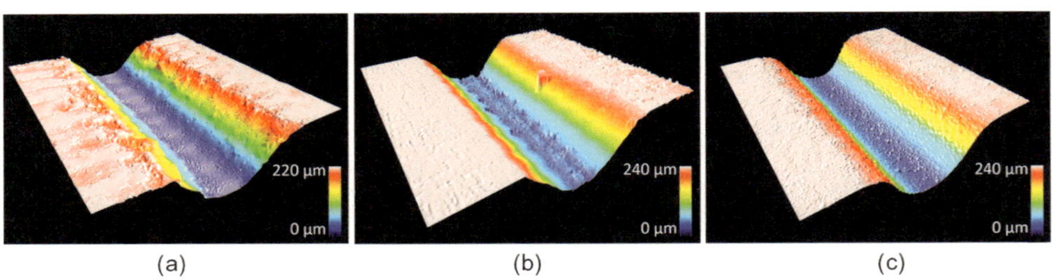

Figure 5.
Confocal images of sections of channels designed with 500 μm in diameter and printed at 45° using: (a) Tough, (b) Clear and (c) Model resin.

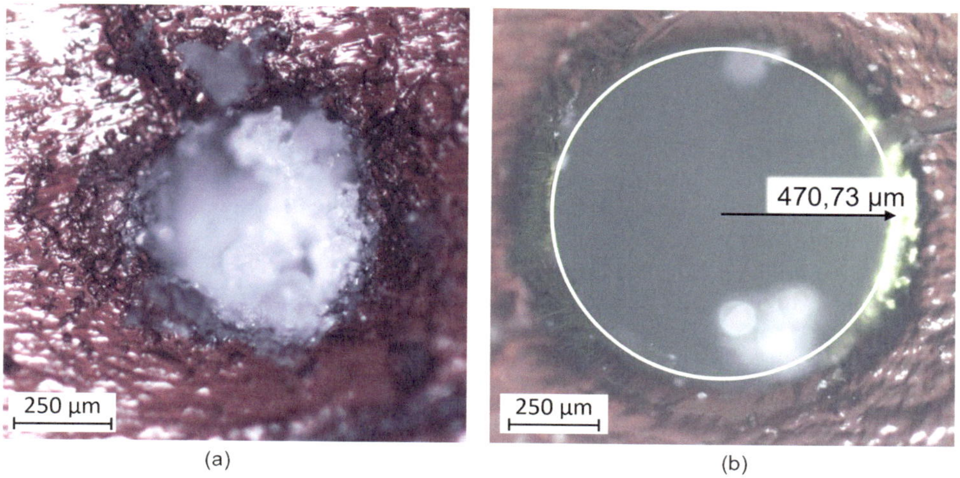

Figure 6.
Microscope images of (a) obstructed and (b) unobstructed channels, with 500 µm of theoretical diameter, printed with amber resin at 0° and 75°, respectively. The images were taken with a 5X microscope objective.

The fact that orientation and diameter are so critical in the manufacture of channels is rooted in the way SLA printers operate and is intimately related to the evacuation of uncured resin, which will be more likely to occur the larger the channel and the more perpendicular the channel to the base (so gravity can enhance evacuation).

4. Conclusions

Microfluidics is a multidisciplinary field that needs versatile technologies capable for manufacturing structures with high accuracy in a precise and reliable way. 3D printing seems to be a promising technology to researchers and industries through easy procedures and a low pollution process. In particular, stereolithographic 3D printers become very attractive due to the developments achieved in lasers, making them one of the most promising choices with greater accuracy and finishing within the existing manufacturing technologies.

The performance in internal channel manufacturing of an SLA 3D printer is tested, since this is one very important piece in several microfluidic devices. Several resins (Clear, Dental, Tough, Amber, Flexible, Elastic and Model) was used for printing the internal channels in terms of accuracy (from hundreds to thousands of micrometres). For this, an annular piece containing several internal channels with different diameters and at different angles was designed and printed for each resin, to analyse the achievable range of dimensions and accuracy.

In light of the results, resin accumulation was found to be the key element behind the correct formation of the channels. This has its origin in the operation principle of SLA printers, based on the layer by layer photopolymerisation of a liquid resin contained in a tank. Thus, the uncured resin must be properly evacuated from the successive layers if a suitable cavity without obstructions and malformations wants to be obtained. It was found that there are two critical parameters: the diameter of the channels and the printing orientation of the device.

While no channel formation was observed for diameters of 250 µm for any of the fabrication angles neither the studied resins, from 500 µm onwards, open lumens began to form. This was the case of Dental and Amber resin, which form channels with printing accuracy (ratio between the printed and theoretical designed diameter) over 80% for values of the angles above 60° and diameters above 500 μm.

In the case of larger diameters (around 1000 μm), the measured accuracies were greater than 70% for every studied resin and grew with the angle. For channels with a diameter of 1500 μm, it was found that all the resins achieved higher accuracy than 90%, so this range can be considered the optimum for the manufacture of complete and fully functional internal channels.

In conclusion, SLA 3D printers are one of the promising technologies in the fabrication of internal channels, showing interesting and promising results for channels of hundreds of micrometres in dimension, very suitable for the growing field of microfluidics. However, the formation of complete internal channels is difficult below 250 μm due to the incomplete evacuation of the uncured resin. There is still room for improvement, and it will be necessary to find both light sources and printing resins that allow higher accuracies, of the order of several tens of micrometres.

Acknowledgements

This work has been sponsored by contracts AEI RTI2018-097063-B-100, AEI/FEDER, UE; ED431B 2020/29; ED431E 2018/08 and ED481D-2021-019, Consellería de Cultura, Educación y Universidad Xunta de Galicia/FEDER e Estructuración Xunta de Galicia, IN607A2019-02. B. Carnero thanks to GAIN/Xunta de Galicia by the contract under no. 11_IN606D_2021_2604925.

Conflict of interest

The authors declare that they have no known competing financial interests or personal relationships that could have appeared to influence the work reported in this chapter.

Author details

Bastián Carnero[1,2], Carmen Bao-Varela[1], Ana Isabel Gómez-Varela[1] and María Teresa Flores-Arias[1]*

1 Photonics4Life Research Group, Applied Physics Department, Instituto de Materiais (iMATUS), Campus Vida, Universidade de Santiago de Compostela, Santiago de Compostela, Spain

2 BFlow S.L., Edificio Emprendia, Santiago de Compostela, Spain

*Address all correspondence to: maite.flores@usc.es

References

[1] Jones N. Science in three dimensions: The print revolution. Nature. 2012; **487**(7405):22-23

[2] Bozkurt Y, Karayel E. 3D printing technology; methods, biomedical applications, future opportunities and trends. Journal of Materials Research and Technology. 2021;**14**:1430-1450. DOI: 10.1016/j.jmrt.2021.07.050

[3] Palmara G, Frascella F, Roppolo I, Chiappone A, Chiadò A. Functional 3D printing: Approaches and bioapplications. Biosensors & Bioelectronics. 2021;**175**:112849

[4] Sandeep B, Kannan TTM, Chandradass J, Ganesan M, John Rajan A. Scope of 3D printing in manufacturing industries—A review. Materials Today: Proceedings. 2021;**45**:6941-6945. DOI: 10.1016/j.matpr.2021.01.394

[5] Bhattacharjee N, Urrios A, Kang S, Folch A. The upcoming 3D-printing revolution in microfluidics. Lab on a Chip. 2016;**16**(10):1720-1742

[6] Nabesawa H, Hitobo T, Wakabayashi S, Aasji T, Abe T, Seki M. Polymer surface morphology control by reactive ion etching for microfluidic devices. Sensors and Actuators B: Chemical. 2008;**132**(2):637-643

[7] Liu RH, Stremler MA, Sharp KV, Olsen MG, Santiago JG, Adrian RJ, et al. Passive mixing in a three-dimensional serpentine microchannel. Journal of Microelectromechanical Systems. 2000;**9**(2):190-197

[8] Lohse M, Heinrich M, Grützner S, Haase A, Ramos I, Salado C, et al. Versatile fabrication method for multiscale hierarchical structured polymer masters using a combination of photo- and nanoimprint lithography. Micro and Nano Engineering. 2021; **10**:100079. DOI: 10.1016/j.mne.2020.100079

[9] Casas-Arozamena C, Otero-Cacho A, Carnero B, Almenglo C, Aymerich M, Alonso-Alconada L, et al. Haemodynamic-dependent arrest of circulating tumour cells at large blood vessel bifurcations as new model for metastasis. Scientific Reports. 2021;**11**(1):23231

[10] Aymerich M, Álvarez E, Bao-Varela C, Moscoso I, González-Juanatey JR, Flores-Arias MT. Laser technique for the fabrication of blood vessels-like models for preclinical studies of pathologies under flow conditions. Biofabrication. 2017;**9**(2):025033. DOI: 10.1088/1758-5090/aa6c3d

[11] Chen YY, Kingston BR, Chan WCW. Transcribing in vivo blood vessel networks into in vitro perfusable microfluidic devices. Advanced Materials Technologies. 2020;**5**:2000103. DOI: 10.1002/admt.202000103

[12] Rupal BS, Garcia EA, Ayranci C, Qureshi AJ. 3D printed 3D-microfluidics: Recent developments and design challenges. Journal of Integrated Design and Process Science. 2019;**22**(1):5-20. DOI: 10.3233/jid-2018-0001

[13] Zhu F, Macdonald NP, Cooper JM, Wlodkowic D. Additive manufacturing of lab-on-a-chip devices: Promises and challenges. Micro/Nano Materials, Devices, and Systems. 2013;**8923**:892344

[14] Romanov V, Samuel R, Chaharlang M, Jafek AR, Frost A, Gale BK. FDM 3D printing of

high-pressure, heat-resistant, transparent microfluidic devices. Analytical Chemistry. 2018;**90**(17):10450-10456. DOI: 10.1021/acs.analchem.8b02356

[15] Cailleaux S, Sanchez-Ballester NM, Gueche YA, Bataille B, Soulairol I. Fused deposition modeling (FDM), the new asset for the production of tailored medicines. Journal of Controlled Release. 2021;**330**:821-841. DOI: 10.1016/j.jconrel.2020.10.056

[16] Heidt B, Rogosic R, Bonni S, Passariello-Jansen J, Dimech D, Lowdon JW, et al. The liberalization of microfluidics: Form 2 benchtop 3D printing as an affordable alternative to established manufacturing methods. Physica Status Solidi. 2020;**217**:1900935. DOI: 10.1002/pssa.201900935

[17] Moreno-Rivas O, Hernández-Velázquez D, Piazza V, Marquez S. Rapid prototyping of microfluidic devices by SL 3D printing and their biocompatibility study for cell culturing. Materials Today: Proceedings. 2019;**13**:436-445. DOI: 10.1016/j.matpr.2019.03.189

[18] Gong H, Bickham BP, Woolley AT, Nordin GP. Custom 3D printer and resin for 18 μm × 20 μm microfluidic flow channels. Lab on a Chip. 2017; **17**(17):2899-2909

[19] Carnero B, Bao-Varela C, Gómez-Varela AI, Álvarez E, Flores-Arias MT. Microfluidic devices manufacturing with a stereolithographic printer for biological applications. Materials Science and Engineering: C. 2021;**129**:112388

[20] Kreß S, Schaller-Ammann R, Feiel J, Priedl J, Kasper C, Egger D. 3D printing of cell culture devices: Assessment and prevention of the cytotoxicity of photopolymers for stereolithography. Materials (Basel). 2020;**13**(13):3011

[21] Hart C, Didier CM, Sommerhage F, Rajaraman S. Biocompatibility of blank, post-processed and coated 3D printed resin structures with electrogenic cells. Biosensors. 2020;**10**(11):152

[22] Xu X, Goyanes A, Trenfield SJ, Diaz-Gomez L, Alvarez-Lorenzo C, Gaisford S, et al. Stereolithography (SLA) 3D printing of a bladder device for intravesical drug delivery. Materials Science and Engineering: C. 2021;**120**:111773. DOI: 10.1016/j.msec.2020.111773

[23] Jacobs PF. Fundamentals of Stereolithography. Valencia, California: Society of Manufacturing Engineers; 1992. pp. 196-211

[24] Lee JH, Prud'homme RK, Aksay IA. Cure depth in photopolymerization: Experiments and theory. Journal of Materials Research. 2001;**16**(12):3536-3544. DOI: 10.1557/JMR.2001.0485

[25] Bennett J. Measuring UV curing parameters of commercial photopolymers used in additive manufacturing. Additive Manufacturing. 2017;**18**:203-212